# HOW ELECTRONIC THINGS WORK . . . AND WHAT TO DO WHEN THEY DON'T

# HOW ELECTRONIC THINGS WORK . . . AND WHAT TO DO WHEN THEY DON'T

ROBERT L. GOODMAN

**McGraw-Hill**

New York  San Francisco  Washington, D.C.  Auckland  Bogotá
Caracas  Lisbon  London  Madrid  Mexico City  Milan
Montreal  New Delhi  San Juan  Singapore
Sydney  Tokyo  Toronto

**McGraw-Hill**

*A Division of The* **McGraw·Hill** *Companies*

1 2 3 4 5 6 7 8 9 0    DOC/DOC    9 0 3 2 1 0 9 8

ISBN 0-07-024630-0

*The sponsoring editor for this book was Scott Grillo, the editing supervisor was Andrew Yoder, and the production supervisor was Sherri Souffrance. It was set in Times New Roman by Lisa M. Mellott through the services of Barry E. Brown (Broker—Editing, Design and Production).*

Printed and bound by R. R. Donnelley & Sons Company.

McGraw-Hill books are available at special quantity discounts to use as premiums and sales promotions, or for use in corporate training programs. For more information, please write to the Director of Special Sales, McGraw-Hill, 11 West 19th Street, New York, NY 10011. Or contact your local bookstore.

This book was printed on acid-free paper.

## DEDICATION

This book is dedicated to my brother, Bill,
for giving me the idea to write this kind of book.

# CONTENTS

# AT A GLANCE

# DETAILED CONTENTS

## Chapter 6    Direct Broadcast Satellite (DBS) system operation

The DBS satellite orbit and tracking details • How the satellites are powered • DBS satellite overview operations • How the RCA DSS system works • Ground station uplink • Data encryption and scrambling • Digital data packets • Operation of DirecTV satellites • How the dish picks up the satellite signal • DBS receiver operations • The receiver modem • Diagnostic test menus • Customer-controlled diagnostics • Signal test • Tuning, phone, and access card test • Overall systems test • Controlled diagnostics, troubleshooting, and service tests • Front-panel control button functions • How to set-up and point the dish • A world view of the total DBS system • The upLink control center • DBS system home view • The telephone jack connections • DBS TV universal remote unit • About the access card • The various DBS receiver front-panel controls and their functions • How to connect your system to a TV and VCR • Different ways to connect your DBS receiver • Some possible system problems and solutions

## Chapter 7    How video cameras and camcorders work

Camcorder features and selections • Digital video images • Video camera and camcorder basics • How video cameras work • What is a camcorder? • Usual camcorder faults • How to localize the camcorder problem area • Mechanical and electronic troubleshooting • Performance check-out • Camcorder functional blocks • Lens system-sync generator circuitry • Camera pick-up devices • CCD pick-up devices • Repairing and cleaning your camcorder • How to remove the cassette lid and deck • Taking apart the lower case section • Cleaning the camcorder heads • Tape will not move and no picture is in the viewfinder • Camera auto-focus operation • Cleaning switches and control buttons • Cassette not loading properly • Intermittent or erratic operation • Camcorder motors • Camcorder troubles and solutions • Tips on caring for your camcorder

## Chapter 8    Wired telephones, cordless phones, answering machines, and cellular phone systems

Telephone system overview • Parts of a conventional telephone • Conventional telephone block diagram and operation • Using the telephone test network box • Some telephone troubles and how to fix them • Static and phone noise checks • Low sound or distortion • DTMF touch-pad problems • Electronic telephone circuit operation • Electronic telephone has noise • No phone operation (dead) • Touch-Tone pad repairs • How an answering machine works • Conventional tape answering-machine operation • Play/record operation • Cassette tape

## Chapter 9    Computer operation, protection, and maintenance

## Chapter 10    Printers, copiers, and fax machine operations

Print cartridge and nozzle operation • Ink-jet head and printer problems and solutions • Some Canon multi-pass troubleshooting tips • Print job vanishes • Machine will not print • Printout is too light • Uninstall program for Windows 95 • Diagnosing software and hardware problems • Plain-paper fax-machine operation • Fax-modem operation and problems • Fax-machine control-panel functions • Some fax problems and solutions • Solutions to fax-machine jammed-paper problems • Dot-matrix printer operation • Block diagram of the dot-matrix printer • Dot-matrix print-head operation • Overall system overview • Dot-matrix printers and print-head troubles and solutions • How laser printers work • Block diagram of the laser printer and operation explanation • Photosensitive drum operation and care • Laser printer problems and tips.

## Glossary

VCR, camcorder, audio tape • Telephone and answering machines • Color TVs and monitors • Computers, printers, copiers, and fax machines

# PREFACE

I think you will find this book very unique in its simple explanations of how electronic equipment works as used in the home or office and its many easy to understand illustrated drawings and photos.

The brain storm for this type book was started many years ago when my brother wanted to know how a picture was formed on a color TV. The planning, development, and portions of the drawings and writing has been in progress for eight years. The actual writing, taking of the many photos and drawings has taken over two years.

The mission of this book is to take the mystery out of how electronic consumer products work for persons with little or no electronic background. Not only does this book give you simplified electronic equipment operations, but hints and tips of what to check when the device does not work properly or not at all. There's also information of how and what to clean, plus preventive maintenance that you can do to extend the life of these very expensive products. It also includes tips on how to protect them from voltage surges and lightning spike damage.

This is a basic "how electronics work" book for the consumer who buys and uses the many wondrous electronic product devices now found in most homes and offices. You now have in your hand a book with over 50 years of my electronic troubleshooting experience and information culled from over 60 of my published electronics books. Thus, this is a book that just about everyone needs to keep on their home or office bookshelf or desk.

The simplified technical electronics information and service tips you obtain from this book can help you in dealing with electronic technicians or service companies when you need professional service for the repairs of your equipment. This might save you repair costs as service personal will not be able to "pull the wool over your eyes," so to speak, because you will be better technically informed. Thus, service repair estimates and costs could swing in your favor. Also, the knowledge gained from this book might help to determine if you should repair the faulty device or purchase a new one.

Finally, this is a valuable book for the hobbyist, electronic experimenter, or any person interested in entering the wonderful world of electronics as a career.

Bob Goodman, CET
Hot Springs Village, AR

# ACKNOWLEDGMENTS

Many thanks to the following electronic companies for furnishing some technical circuit operation information, drawings, and photos:

Zenith Electronics, Thomson Consumer Electronics, Sencore Electronics, and Bose Acoustic Wave Music System

A very special thank you to Ms. Frankie Rundell who performed word-processor operations and corrections for this book's manuscript.

# INTRODUCTION

Chapter One of this unique book gives you a very basic introduction to electronics. Thus, the reason I named it "Very Basic Electronics 101".

This chapter will contain photo and drawings of the components found in your electronic devices with explanations of what the components do, how they are constructed and how to test them.

You will be shown how to use a volt-ohm or multimeter to check voltage and resistance found in electronic circuits. Also, how to use the volt-ohm meter to check circuit boards and components. And even a simple circuit tester you can build in order to check solid state devices such as transistors, diodes and IC's. This is referred to as a component curve tracer. The chapter ends with how to use a soldering iron and techniques for soldering in components and safely de-soldering electronic components.

Chapter Two has an overview of how FM radio signals are developed and then received on an FM stereo radio. Also, a look at the "DOLBY" audio system and radio troubles and repair tips. Then you will see how loud speakers work and the advanced "BOSE-ACOUSTICS" radio and speaker systems operation.

The chapter concludes with how cassette recorder/player machines work, audio cassette trouble symptoms, corrective action and care and cleaning of these units.

Chapter Three let's you discover how the audio and video laser disc players/compact disk (CDs) operate and how to clean them plus perform minor service repairs.

Next will be tips on how to clean the laser pick-up, mechanical disc drive and the disc slide operations. Also, hints on keeping your CD operating smoothly plus common CD problems and solutions.

The last portion of the chapter covers various types of remote control units for TV sets, CD players, cable control boxes, VCRs and DSS satellite dish receivers. And of course some simple checks and repairs of these remote units.

Chapter Four begins with an overview of the color TV signal makeup, the components within this signal and some of the various world wide color TV standards.

Then onto the color TV block diagram which will help walk you through the circuit operations within these blocks. These various stages that make up the color TV sets operation are then explained in more detail.

Horizontal and vertical sweep circuit operations are now delved into and what problems to look for in color TV sets and computer monitors. Some actual troubles and solutions are also investigated.

Color picture tube operation is next on the agenda and how it develops a color picture on the screen with its associated high voltage supply, problems and solutions. This is followed by large screen projection TV set operation details.

This chapter concludes with various color TV and PC computer monitor trouble symptoms and their solutions.

Chapter Five tells about the operation of a machine that can store TV video information and sound on a magnetic tape for later playback. This, of course, is the HOME VIDEO CASSETTE RECORDER or VCR machine. This chapter then gives you information on how to clean the VCR video heads, other parts of the machine, routine VCR maintenance and adjustments. The last portion has some symptom/problems that may occur with VCRs and gives you tips on how to correct them.

Chapter Six has information on the new and exciting DIGITAL TV DIRECT SATELLITE (DSS) transmission system and its operation. An overview follows of the system that includes the up-link earth stations, the satellite that receives and retransmits the signals and dish/receiver that picks up the down link signals. Now, onto your dish/satellite receiver operations and trouble hints and solutions. And then some signal data encryption and scrambling to eliminate any unauthorized (free) TV viewing.

The chapter concludes with a "World View" of the total DSS system operation and features. And then some drawings that shows you how to connect the DSS receiver to your TV set and VCR units.

Chapter Seven commences with a look at the past and present video cameras and camcorders that have been available. We will review various camcorder features, older ones with vidicon pick-up tubes and the modern CCD solid-state image pick-up chips and digital video cameras. How the camcorder works will be explained and how to make minor repairs and clean the recording heads. The chapter ends with repair techniques and probable camcorder problems, their solutions and ways to solve them. A list of camcorder and video, camera troubles, solutions and how to repair them rounds out this chapter.

Chapter Eight starts with an overview of a telephone land line system and home phone operation. We then show you how the electronic phone works and ways to determine if your phone or the phone company line is at fault. Then some tips on phone repairs and cleaning.

The chapter continues with how answering machines work and their cleaning, repair techniques. Cordless phone operation is explained next and the radio frequenices that are available for them. Interference, noise problems, range and cordless phone privacy is then discussed. All types of phone problems are given along with how to repair and clean them.

The chapter concludes with how two-way radio trunking systems and cellular phones systems operate. Then some battery tips and cellular phone reception problems that you may encounter are explained.

Chapter Nine delves into the inside working's of your personal computer (PC), how you can protect it from power line surges, perform minor repairs and clean the various sections of the computer. You will see how each basic block works and then how they all operate together to complete the total system. You will have an insiders view of the popular INTEL Pentium microprocessor found in many PC's and see how the hard drive, floppy disc and CD ROMM drive works.

The chapter continues with the operation of the PC keyboard and how the mouse is used to control your PC functions. The Modem operation is explained along with the workings of the Internet Super Information Highway.

The last portion deals with problems that various sections of your PC may have plus tips and solutions on how to solve them. Information on diagnostic software disks to locate and fix various program faults and viruses.

Chapter Ten begins with a brief overview of basic printer, copier and FAX machine operations. You will find out how the "Daisywheel" printer, Ink-Jet printer, Dot-Matrix, and Laser printers operate. Also information on copiers and FAX machine operations. There will also be troubleshooting information, problems and solutions and care and clenaing for all of the printers covered in this chapter.

Electronic Glossary Terms has been included to help you better understand some of the terms used in explaining "How Electronic Things Work" in the previous book chapters.

# INTRODUCTION TO VERY BASIC ELECTRONICS "101"

# How Resistors Work

Resistors are made in various shapes, sizes, resistance values (in ohms) and wattage ratings. Resistors are the most common electronic circuits. In fact, ICs have many resistors inside them. Resistors are used as current-limiting devices and an electronic circuit will not work without them. You might think of a resistor as a control device that limits current flow to the circuit load. A circuit load provides the work; it can be a light bulb, motor, loud speaker, transistor, or IC. Resistor values are in ohms and are made of carbon or coils of resistance wire. Resistor values can be fixed or adjustable (as with a rheostat or like a variable volume control used on a radio or TV). The value in ohms of a resistor is what will determine the electron current. A low resistance will cause a large current to flow, and a high resistance will cause a small current flow.

## RESISTOR TYPES

Many types, values, and sizes of resistors are used in electronic products. The photo in (Fig. 1-1) shows various wattage sizes of carbon and flameproof resistors from ¼-watt to 2-watt ratings. These fixed resistors are made to a specific resistance value and cannot be changed. The resistance value is indicated by color-coded bands or stamped numbers on the side of the resistors body. The symbol for a fixed resistor is shown in Fig. 1-2. The larger, 10- to 300-watt, power resistors are shown in Fig. 1-3.

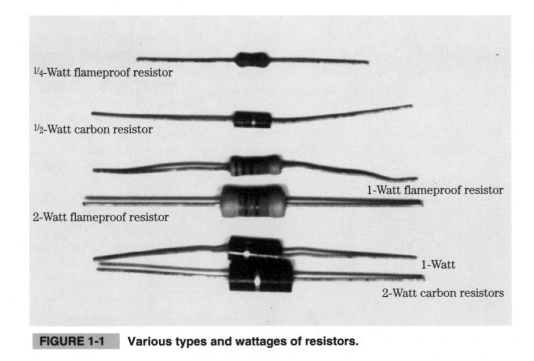

¼-Watt flameproof resistor

½-Watt carbon resistor

1-Watt flameproof resistor

2-Watt flameproof resistor

1-Watt

2-Watt carbon resistors

**FIGURE 1-1    Various types and wattages of resistors.**

Carbon Resistor

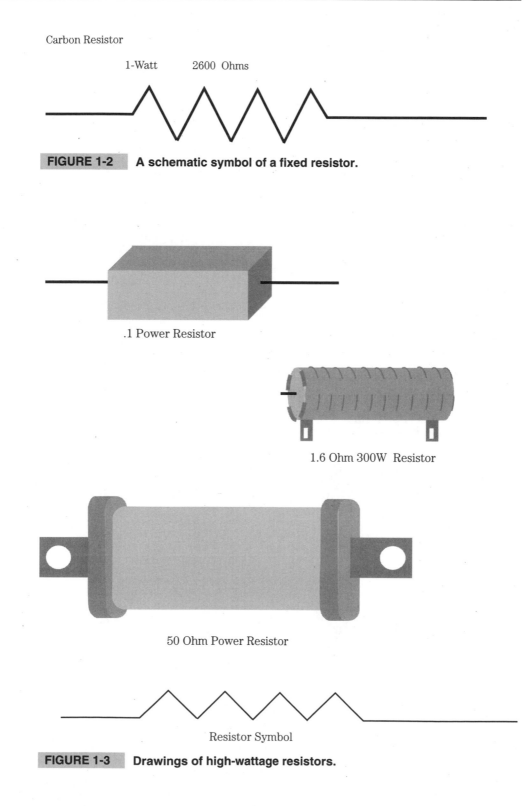

1-Watt        2600  Ohms

FIGURE 1-2    **A schematic symbol of a fixed resistor.**

.1 Power Resistor

1.6 Ohm 300W  Resistor

50 Ohm Power Resistor

Resistor Symbol

FIGURE 1-3    **Drawings of high-wattage resistors.**

## READING RESISTOR COLOR CODES

If you need to replace a resistor, you will need to be able to read the color code bands to determine its value because you might not have a schematic or the value might not be given on the circuit diagram. The standard resistor color code is:

Black    0
Brown    1
Red      2
Orange   3
Yellow   4
Green    5
Blue     6
Violet   7
Gray     8
White    9

Most fixed carbon resistors use the color band layout (as shown in Fig. 1-4) to indicate their value and tolerance. The first band color is for the first number of the resistor value. Band 2 indicates the second number. Band 3 is a multiplier to show how many zeros follow the first two color-band numbers. As an example, a 25,000 ohm (25 kΩ) resistor would have these band colors:

■ Band #1: Red or 2
■ Band #2: Green or 5
■ Band #3: Orange or 3 for 3 zeros 000.

And the resistor would read as 25,000 ohms.

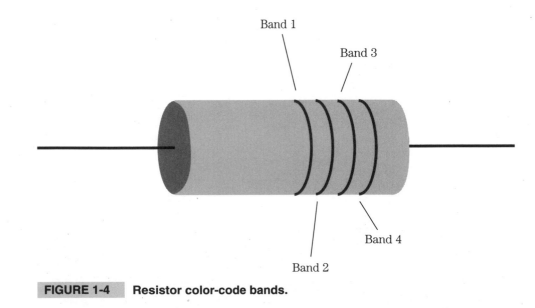

**FIGURE 1-4**    Resistor color-code bands.

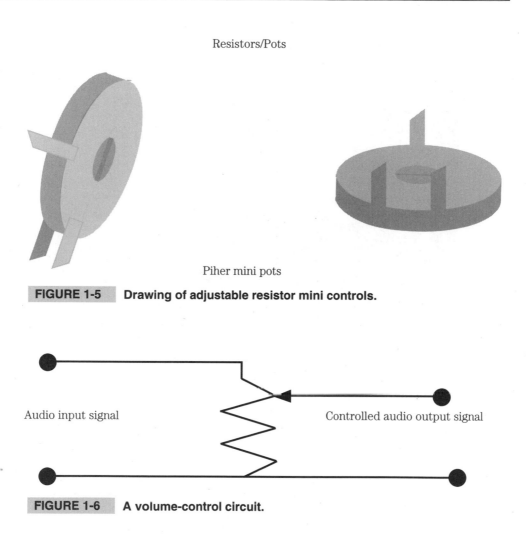

Resistors/Pots

Piher mini pots

**FIGURE 1-5**    **Drawing of adjustable resistor mini controls.**

Audio input signal

Controlled audio output signal

**FIGURE 1-6**    **A volume-control circuit.**

The fourth band is reserved for the tolerance band color. A silver color band shows that the resistance value indicated is within ±10 percent. If the band is gold, the tolerance is 5 percent of the stated value.

**Variable resistors**   A variable resistor that you can rotate or slide is called a *potentiometer*. These are used for radio and TV volume controls or circuit adjustment controls, as shown in Fig. 1-5. The potentiometer is used in circuits where the voltage needs to be controlled from zero to the maximum, then back to zero. Figure 1-6 illustrates how a volume control would appear in a circuit schematic.

The small volt-ohm meter shown in Fig. 1-7 can be used to check fixed resistors for their correct value or opens, and also to check variable resistor controls for a bad spot when it is rotated. Sometimes these variable volume controls can be cleaned with a spray control cleaner and restored to proper operation.

**FIGURE 1-7    A small volt/ohm multimeter that is used to check resistance values.**

## RESISTOR PROBLEMS

A resistor might be defective because of a manufacturing fault or even years of use in a high-moisture area. But, the great majority of resistor failures are caused by circuit faults or lightning/power surges. If you find a burnt resistor, it will probably be caused by one of the following reasons:

- A short on the circuit board or wiring, which might have had a liquid spilled onto it.
- A shorted or very leaky capacitor.
- A shorted transistor, diode, or IC.
- A power-line surge or lightning spike damage.

If a carbon resistor has been overheating slowly for a long time, the resistance will be lower in value. If it has burned rapidly because of a short circuit, the resistor might go up in a puff of smoke and the resistor will be open. In fact, you might only see a black, burnt area with two wire leads sticking up. An intermittent resistor is a rarity and usually is of the wirewound variety. The intermittent ones will usually look normal, but if you see one that has a crack in it, replace it.

Always replace resistors with the same type and value. You can replace a resistor with one that has a larger wattage rating, if you have room to mount it on the circuit board.

# Electronic Circuit-Protection Devices (Fuses)

When electronic equipment fails, depending on the fault, the power supply circuit will draw more current. To protect the other circuits from more damage, a fuse is installed to shut down the device. The fuse is placed in series with the current-drawing circuit, as indicated in Fig. 1-8. A blown fuse is, in effect, the same as turning off the power switch.

Many different types of fuses are used in consumer electronic devices. Figure 1-9 shows four types of fuses. Some of the different types of fuses used for electronic circuit protection are:

- Very small size microfuses.
- Fast-acting or quick-blow glass fuses.
- Slow-blow or lag-time fuses.
- Ceramic fuses.
- Slow-blowing glass fuses.

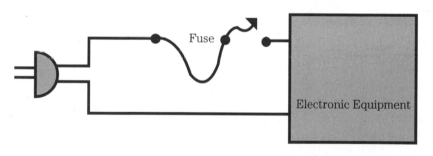

AC Power line (fused) operated equipment

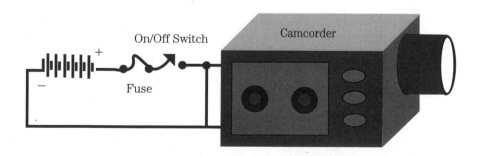

Fuse placement for battery operated equipment.

**FIGURE 1-8**    **Where protection fuses are installed for ac power line and battery operated electronic devices.**

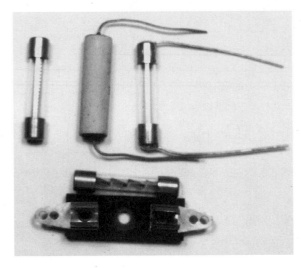

**FIGURE 1-9**    Various types of fuses used in consumer electronic devices. From the top left to right is a glass slow-blow fuse, a ceramic fuse, and a glass fuse with leads attached (pig tails) that can be soldered into a circuit board. At the bottom is a fast-blow fuse that can be snapped into a fuse holder.

- Thermal-protection fuses.
- Fusible resistor-type fuses.
- Bimetal strip circuit breakers that can be reset.

## TESTING THE FUSE

A fuse is either good or bad. With some glass fuses, you can see that the link has been blown or is missing. However, with other types of fuses, you cannot see inside or you cannot be sure if it is good or bad. To be sure, use an ohmmeter, as shown in Fig. 1-10, to

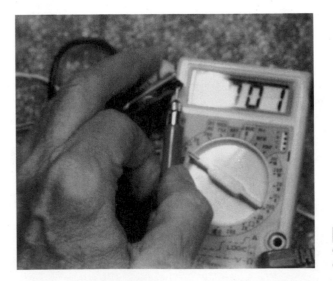

**FIGURE 1-10**    A digital ohmmeter being used to check a fuse.

check for continuity. The ohmmeter will quickly indicate if the fuse is good or bad. Usually, if the fuse has blown, the circuit is shorted or a high current is being drawn. Some fuses might fail from a defect, vibration link breakage or if loose in the fuse holder, might become very hot and will actually melt the solder alloy link and cause the fuse to be open. Always replace a fuse with one that has the same value.

# How Capacitors Work

Capacitors are used in electronic circuits for isolation or blocking dc voltages, and used with coils to produce resonant or tuned circuits, transfer of ac signals, filtering out unwanted interference, and as smoothing filters in power supplies.

The construction of a basic capacitor is shown in Fig. 1-11. It consists of two plates (conductors) that are separated by an dielectric (insulator). The insulator material can be mica, paper, tantalum, plastic, fiber, or even air.

The capacity in (microfarad or picofarad) determines what size of electrical charge that the capacitor can store (hold). The capacity is determined by the size of the plates, the space between the plates, and the type of dielectric between the plates.

## TYPE OF CAPACITORS

Figure 1-12 shows some disc ceramic capacitors on the right side and tubular electrolytic capacitors on the left side. Figure 1-13 shows a variety of teflon epoxy dipped capacitors. Figure 1-14 shows an assortment of mica trimmer capacitors and also several mini adjustable trimmer capacitors.

The names of various types of capacitors are as follows:

■ Can, tubular, and molded electrolytic capacitors are used for power-supply filtering.
■ High-voltage ceramic capacitors.

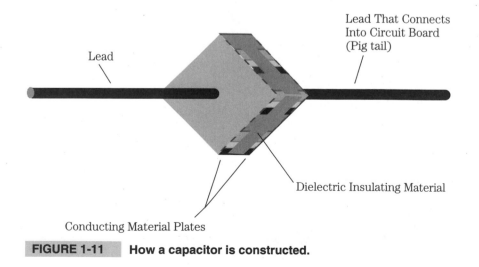

**FIGURE 1-11**    **How a capacitor is constructed.**

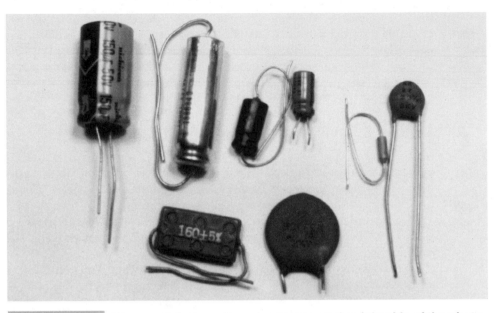

**FIGURE 1-12**    Disc ceramic capacitors are shown on the right side of the photo and tubular electrolytic capacitors are shown on the left side.

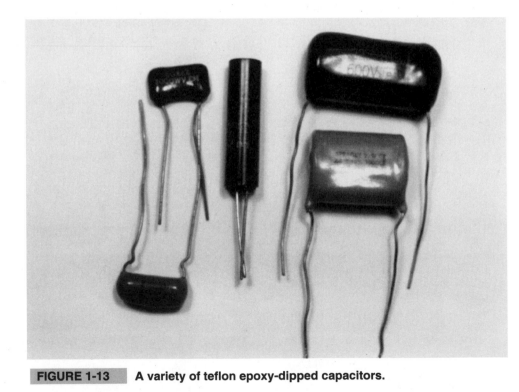

**FIGURE 1-13**    A variety of teflon epoxy-dipped capacitors.

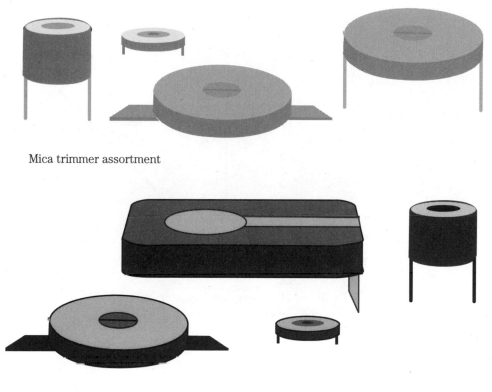

Mica trimmer assortment

Mini trimmers

**FIGURE 1-14**     **An assortment of mini mica trimmer capacitors.**

- Tubular ceramic capacitors.
- Feed-through capacitors.
- Disc ceramic capacitors.
- Variable air tuning capacitors.
- Adjustable trimmer (mica and ceramic) capacitors.

## CAPACITOR CIRCUIT DIAGRAM SYMBOLS

Figure 1-15 shows the circuit schematic symbols for a electrolytic filter capacitor, fixed capacitor and a variable capacitor. Figure 1-16 depicts electrolytic filter capacitors being used in a power-supply circuit. The tuned circuit shown in Fig. 1-17 shows a variable air capacitor with a small trimmer capacitor across it, as used in a radio to select the various station frequencies. A shaft is connected to the plates of the rotor, which rotate to change frequencies. The plate area is either decreased or increased, which, in turn, changes the capacitance value, and thus the frequency of the tuned circuit.

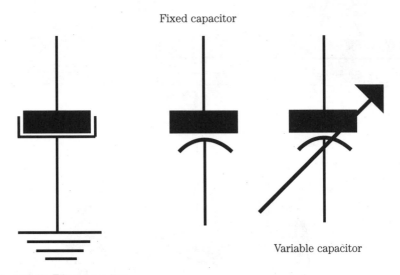

Fixed capacitor

Variable capacitor

Electrolytic Filter capacitor

**FIGURE 1-15**    Schematic circuit drawings of various types of fixed and variable capacitors.

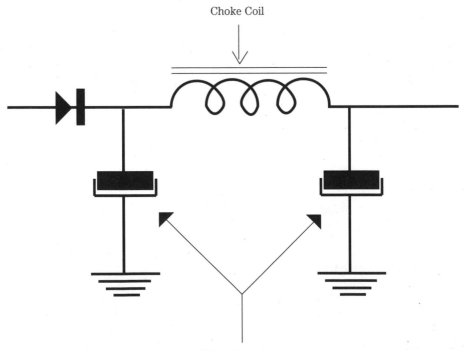

Choke Coil

Filter Capacitors

**FIGURE 1-16**    Filter capacitors are used in a power supply to smooth out the rectified dc pulse voltage.

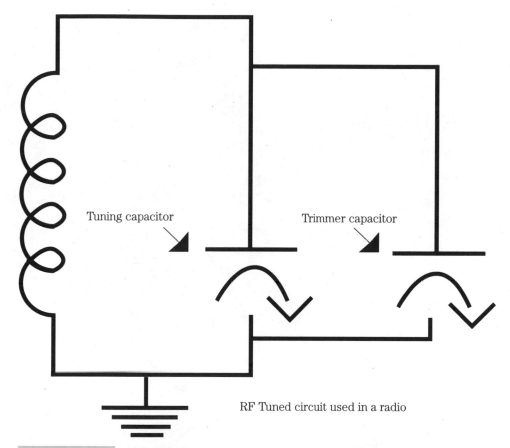

Tuning capacitor

Trimmer capacitor

RF Tuned circuit used in a radio

**FIGURE 1-17**    **A variable tuning capacitor circuit with a trimmer in parallel with it. This circuit is used to tune a radio to various frequencies.**

## TIPS FOR LOCATING FAULTY CAPACITORS

When a capacitor fails, it might become shorted, open, or leak. A capacitor checker can be used to find these problems, but this equipment is expensive. However, you can use a volt/ohm multimeter, like those shown in Fig. 1-18. You can use the ohmmeter range to see if the capacitor is shorted or open. If you measure a low-resistance reading, the capacitor is shorted or very leaky. If you obtain a high resistance reading with a "kick" of the meter needle, the capacitor is probably good. For an accurate, reliable test, the capacitor should be removed from the circuit.

In any circuits that contain solid-state devices (diodes, transistors, ICs, etc.), do not bridge a good capacitor across a suspected faulty one for a test. A spark will usually occur, which can damage the junctions within other solid-state devices mounted on the PC board.

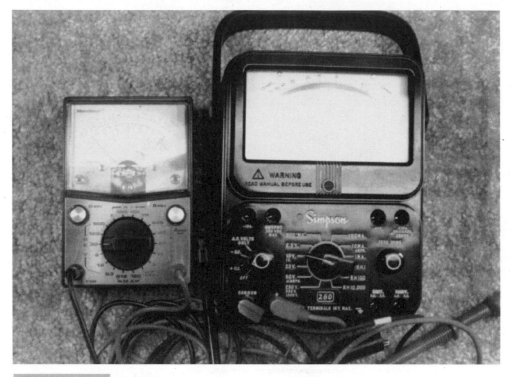

**FIGURE 1-18**    Volt/ohmmeters that can be used to check capacitors and other electronic components.

# Transformer and Coil Operations

Transformers and coils are used in most electronic consumer devices. These can be power transformers in the power-supply section, radio-frequency (RF) transformers and coils in the RF and IF sections of TV and radio receivers, and chokes or coils used to eliminate various types of RF and electrical interference.

Figure 1-19 shows schematic symbols of transformers and choke coils that use iron, ferrite and air for the core forms. Transformers will usually have four or more leads and choke coils will only have two lead wires for connections. As the name implies, the transformer "transforms" pulsing dc or ac voltage up (to a higher voltage) or down (to a lower voltage) by induction from one winding to another adjacent near-by winding(s).

Transformer action can only occur when the voltage to the coil winding is changing, such as an alternating ac (alternating current) voltage. If a dc voltage is connected to a transformer winding primary, the secondary winding would only produce a voltage pulse for an instant when the input coil voltage is connected or disconnected. The magnetic field produced by the primary coil will (cut) go across the secondary winding and by this magnetic induction will induce an ac voltage into the secondary coil. Thus, a magnetic transfer occurs, which increase or decrease the ac output voltage, depending on the number of turns of the coil.

Transformer types used in electronic equipment are:

- Power transformers.
- Audio transformers.
- Voltage regulator and smoothing transformers.
- Antenna matching transformers.
- Oscillator transformers.
- Adjustable core (slug) transformers.
- Radio-frequency (RF) transformers.
- High-voltage "flyback" transformers used in TVs
- IF transformers.
- Interstage audio-matching transformers used in audio amplifiers.

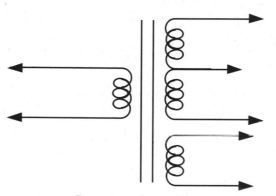

Power transformer ( iron core )

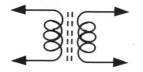

Ferrite-core transformer

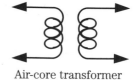

Air-core transformer

Air -core choke

Ferrite-core choke

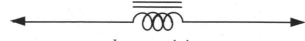

Iron -core choke

**FIGURE 1-19**    Schematic symbols for transformers and choke coils used in
electronic equipment.

Power transformer

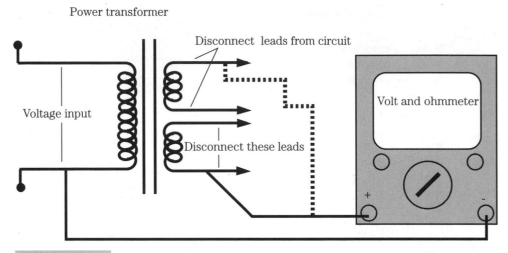

**FIGURE 1-20**    **How to use an ohmmeter or voltmeter to check for a short or leakage between windings of a power transformer.**

## TRANSFORMER TROUBLES AND CHECKS

Transformers can fail in many ways. Some of the various failures are:

- The coil wire turns can open. This can occur where the copper wire coils are connected to the terminal lugs.
- The windings can become shorted to adjacent turns of the coil wires.
- The primary and secondary coil windings can become shorted to each other.
- The primary and secondary coil windings can develop a high-resistance leakage.
- The coil windings can become shorted due to insulation breakdown to the metal core, transformer case, or frame.
- A power transformer might become hot, have a waxy material start leaking from the case and might actually smoke and burn up.

If you detect a burning odor from your equipment and see some melted wax coming from inside the transformer case located in the power-supply section, immediately turn it off or unplug the device. You might find the same symptoms if your equipment stops working and a fuse has blown. An overheated power transformer will usually not be damaged as the problem that caused this condition is some other component that has shorted out. These component faults could be a shorted diode rectifier, electrolytic filter capacitor, regulator transistor or a bypass capacitor. You can use an ohmmeter to check for any shorts or low resistance in the B+ supply lines.

Figure 1-20 illustrates how you can check for leakage between the primary and secondary of a transformer. With the device turned on and a dc voltage on the primary winding, measure for any voltage with your voltmeter at the points indicated on the two secondary windings. If you find even a very small voltage, the transformer has leakage and should be replaced. The ohmmeter is used to check across each winding for opens. Be sure that the device is turned off or unplugged for these ohmmeter checks.

Because of the high-voltage involved with the high-voltage sweep or "flyback" transformers in color TV sets and monitors, they usually are the cause of the failure when they arc, smoke, or burn up.

# Transistors, Integrated Circuits (ICs), and Diodes

This section shows how diodes, transistors, and ICs work, what they look like in your equipment, and some ways to check them out for failures. Figure 1-21 shows a 24-pin IC used in some camcorders and are sometimes called *microchips*, *chips*, or *ICs*. An IC consists of many solid-state transistors, diodes, resistors, coils, and capacitors. You can think of the transistor as the basic building block of which all IC chips are constructed. When ICs are used in computers, many transistor gates create binary data to either be: "off" or "on," which provide "0s" and "1s." Also, transistors make it possible to use a very small electrical current to control a much stronger second current. Transistors are called *semiconductor devices* (hence, solid-state) because they are actually made from materials, such as silicon and germanium, which are not perfect insulators nor good conductors. So, the current in these solid-state devices is controlled within a "solid-state" material.

## DIODES

The diode is a solid-state device that will only permit current flow in one direction, or polarity, but not going in the opposite direction. It can be used as a protection device for dc-operated equipment. If equipment is connected accidentally to a battery with the wrong polarity, no damage would occur because no current would flow.

Diodes are very useful in power supplies to change ac voltage into pulsating dc voltage when used as a rectifier diode. The top drawing of Fig. 1-22 illustrates how the current will flow in one direction only through the diode. The bottom drawing of Fig. 1-22 is of a simple power supply, where the diode is used as a rectifier diode to change an ac voltage into a pulsating dc voltage, which is then smoothed out with filter capacitors for a dc voltage. Figure 1-23 shows the many various shapes and sizes of some common diodes in consumer electronic equipment.

With the correct polarity the voltage across a diode will let the current pass with no resistance or very easily. With the opposite polarity of voltage, the current will encounter a very high resistance and current will not flow. When the current cannot pass through the

**FIGURE 1-21**   **A 24-pin IC used in a camcorder.**

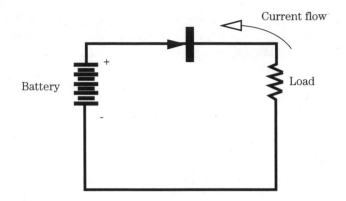

A diode will only let current flow in one direction

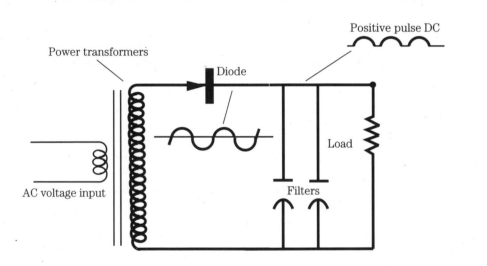

How a diode is used in a power supply circuit the diode changes AC voltage to a DC voltage.

**FIGURE 1-22**    **The top drawing shows how a diode will only let current flow in one direction. The bottom drawing illustrates how a diode is used as a rectifier in a power-supply circuit.**

diode, this is called *reverse bias*, but, when the current can easily flow through the diode, this is referred to as *forward bias*.

## METAL-OXIDE VARISTOR (MOV) OPERATION

The MOV or varistor is used in many consumer electronic products. Figure 1-24 depicts the MOV or varistor in circuit diagrams. You can think of the varistor (voltage-variable resistor) as a device that has a high resistance at a low voltage, but with a certain higher voltage, the resistance drops to a much lower value.

MOVs usually consists of a zinc-oxide material. Because the varistor is not polarized, it is very useful in ac circuits. MOVs are used across the ac power line into electronic equipment and also in the telephone line input circuits (Fig. 1-25). The MOV is usually installed

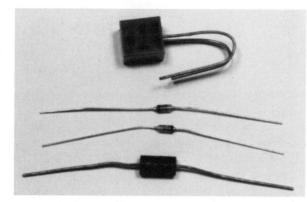

**FIGURE 1-23**   Some various shapes and sizes of diodes used in consumer electronic equipment.

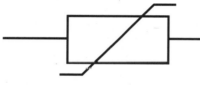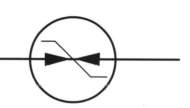

**FIGURE 1-24**   The schematic symbols for an MOV or varistor.

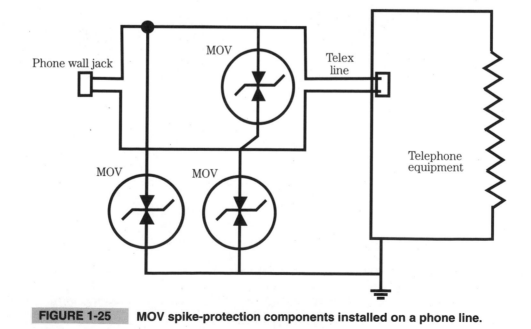

**FIGURE 1-25**   MOV spike-protection components installed on a phone line.

for spike and voltage-surge suppression for circuit protection. The MOV is rated by its "breakdown" voltage, thus when installing or replacing an MOV, be sure that the rated voltage is a little higher than the voltage usually found at this circuit point.

# How Transistors and ICs (Solid-State Devices) Work

You will find many transistors used in all electronic equipment. And, of course, integrated circuits (ICs) have lots of transistors inside their chips. The transistor package has three leads coming out (sometimes four leads) and is a solid-state electronics package that can perform amplification and switching of electronic signals. Figure 1-26 shows the various types and sizes of transistors used in consumer electronic products. There are two basic transistor designs. One type is the *bipolar* and the other is the *Metal-Oxide-Semiconductor Field Effect (MOSFET)*. Figure 1-27 illustrates a cross-section view of a NPN bipolar transistor structure. A bipolar transistor is usually made of a silicon material. Discrete transistor construction requires many complex steps that start with a blank wafer of silicon. Some of the steps include photographic masking, photo reduction of large-scale artwork, ultraviolet light to alter the chemical composition, and chemical solvent to remove unexposed photoresist.

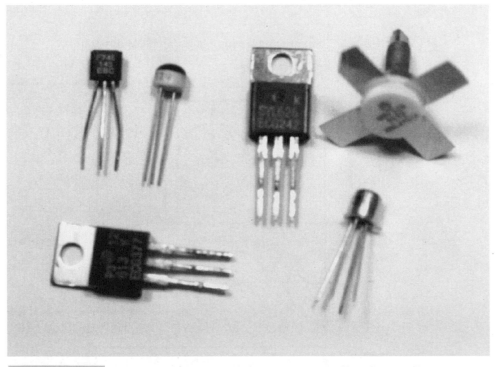

**FIGURE 1-26**    Various transistors used in consumer electronic products.

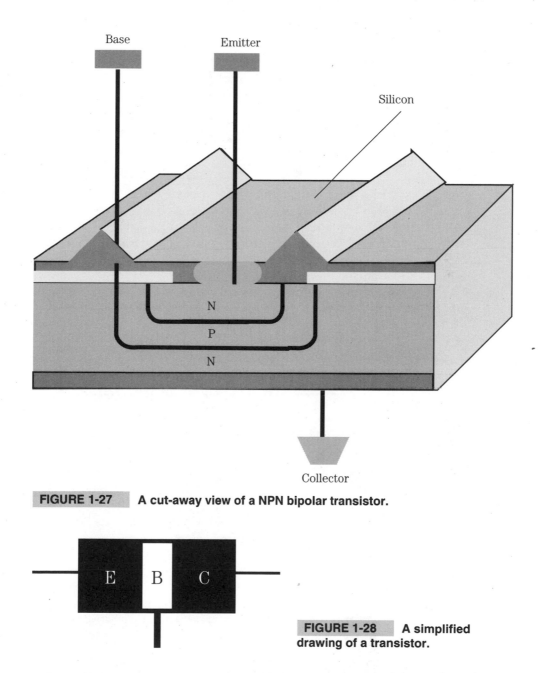

**FIGURE 1-27**    **A cut-away view of a NPN bipolar transistor.**

**FIGURE 1-28**    **A simplified drawing of a transistor.**

A transistor consists of thin layers of material with a collector on one side, a thin base layer in the middle, and the emitter on the other side. Notice the simplified drawing in Fig. 1-28. The material used for the emitter and collector sections are opposite of that used for the base.

For a PNP transistor, N-type material is used for the base, but the collector and emitter are made from P-type material. With an NPN transistor, the base is an P-type material and

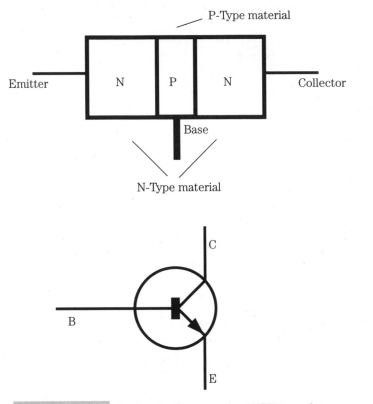

**FIGURE 1-29**    **A circuit diagram of an NPN transistor.**

the collector and emitter are made of N-type material. A drawing of an NPN transistor with the circuit diagram below it is shown in Fig. 1-29. Figure 1-30 shows a drawing of a PNP transistor with its circuit diagram below it.

Review this one more time. With a NPN transistor, the base is a P-type material and the collector and emitter are both made from N-type material. Conversely, for a PNP transistor, the base is an N-type material, and the collector and emitter are of the P-type material.

In most cases, the NPN and PNP transistors work the same way in their circuits, except for the applied voltage polarities. A PNP transistor will have a negative collector voltage and a negative base bias voltage. An NPN transistor will use a positive voltage on the collector and a positive bias, thus it has a collector-to-emitter positive voltage.

## THE INTEGRATED CIRCUIT (IC)

Inside an IC package is a small "chip" or microcircuit with many active and passive electronic parts interconnected on a small semiconductor substrate or wafer. The chip will perform many electronic circuit functions in a very small space. Figure 1-31 illustrates the many transistors, resistors, and capacitors found in a typical op-amp IC package.

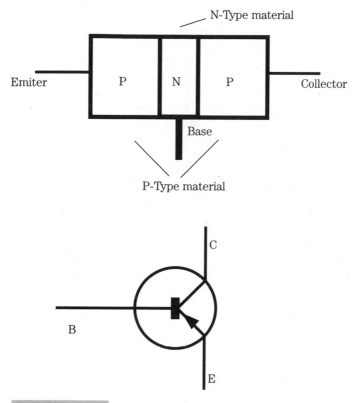

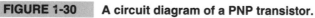

**FIGURE 1-30**    A circuit diagram of a PNP transistor.

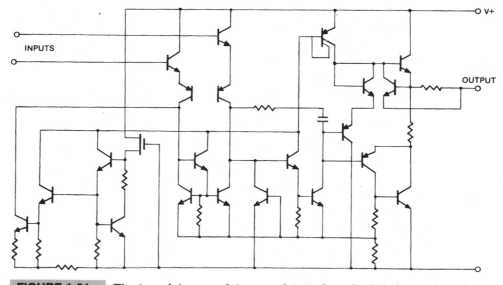

**FIGURE 1-31**    The transistors, resistors, and capacitors inside of a typical IC op amp.

Some of the advantages of ICs over conventional circuits are:

- Because all circuit parts are on the same substrate, performance and temperature conditions will not vary.
- Because of the microchip construction, more circuit operations can be mounted on smaller circuit boards.
- The IC is the main reason that electronic products are more reliable because so many external electrical connections have been eliminated.
- The IC has increased circuit performance and speed because of shorter lead interconnections. The invention of the IC caused the "great leap forward," which made possible the increased speed of computer computations and vast amounts of memory retention.
- With lower power consumption and less heat loss, the IC has made electronic devices much more efficient.

The circuit diagrams for two types of ICs are shown in Fig. 1-32 and are the way you will find chips drawn on schematic diagrams. A photo of the round 8-pin IC is seen in Fig. 1-33 and its circuit drawing is shown on the right side of Fig. 1-32. A photo of some common 16- and 18-pin in-line ICs are displayed in Fig. 1-34.

# Solid-State Scope Sweep Checker

You can build this simple checker that tests transistor, diode, Zener diode, SCRs, and even some ICs. This device connects to an oscilloscope to make a fast "go-no-go" test unit. This

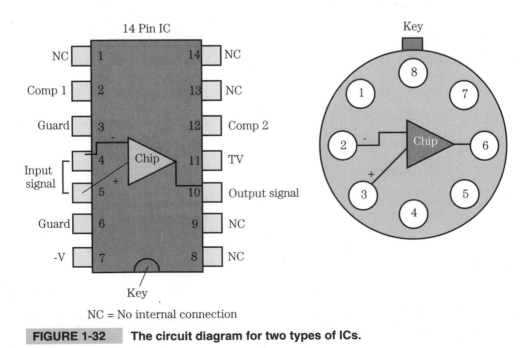

NC = No internal connection

**FIGURE 1-32**    **The circuit diagram for two types of ICs.**

**FIGURE 1-33**  An 8-pin plug-in IC.

**FIGURE 1-34**  Some 16- and 18-pin in-line ICs.

little sweep checker can even be used to check resistors, capacitors, and find shorts and open circuits. I designed this "sweep junction" checker over 30 years ago; for some time, Texas Instruments (TI) used this device to sort out defective transistors and diodes on their production line.

This simple checker is easy to build-up and connect to your scope. You should find it to be a quick and easy, but reliable, test box for fast checking most solid-state devices. This tester, which connects to a scope, uses an ac sinewave to sweep the solid-state devices junction under test.

The simple circuit diagram for this checker is shown in Fig. 1-35 and also how to connect it to the scope's vertical and horizontal sweep inputs. Transformer T1 has a 120-Vac primary and either a 6.3-V or 12.6-Vac secondary. You can use red and black voltmeter leads with needle point tips for the test probes. The black lead is ground and the red lead is used for the positive test lead. The polarity of the leads will affect the scope waveforms by flipping the trace upside down when you reverse the leads to the component under test. The six scope pattern drawings shown in Fig. 1-36 are some typical traces you will find for the various components listed.

When checking a solid-state device out of the circuit, the main point of interest is the knee of the curve. A sharp bend usually indicates that the device is good. A straight horizontal

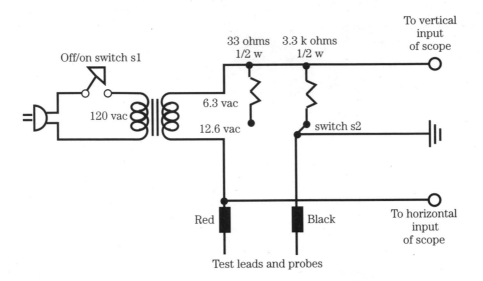

Connect test probes across solid-state device to be checked with no power
applied to circuit under test.

**FIGURE 1-35**    A circuit diagram for a solid-state sweep/junction checker.

Good junction waveform

Shorted component

An open circuit waveform

Zener junction waveform

Capacitor produce
an oval waveform

Some resistence.  The
angle changes with ohms
value

**FIGURE 1-36**    Some typical scope waveforms that you will find when using
the junction sweep checker.

line indicates an open junction and a straight vertical line on the scope pattern means a shorted
component. If the supply voltage of the curve checker exceed the peak-inverse voltage (PIV)
of the solid-state junction under test, Zener action might occur. This is indicated by a very
short vertical line, see Fig. 1-37, at one end of the trace pattern and should be disregarded.

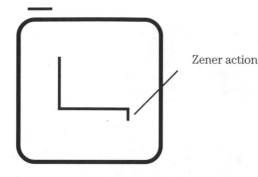

Zener action

**FIGURE 1-37**   **The short line at end of the trace will occur when curve tracer voltage exceeds the PIV of the solid-state junction under test. Just disregard this line.**

When solid-state devices are checked in circuit, the ideal out-of-circuit scope traces might not appear because other resistors, coils, and capacitors in the circuit might cause the trace patterns to vary. Thus, when checking in-circuit components, a comparative method must be used. Also, for a positive test, the component can be removed from the circuit.

When checking transistors, disconnect power from the device under test and connect the test probes. Always connect the test probes to the transistor terminals by the color code shown:

- Base-emitter junction: Base red, emitter black.
- Base-collector junction: Base red, collector black.
- Collector-emitter junction: Collector red, emitter black.

The shape of the pattern shown in Fig. 1-38 is for a transistor with a high junction leakage. The pattern in Fig. 1-39 was obtained when the circuit under test had a very low resistance value.

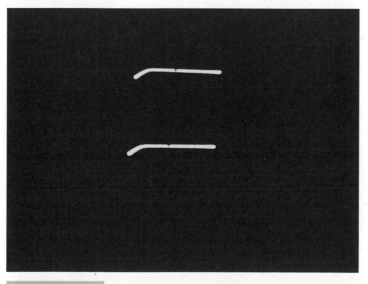

**FIGURE 1-38**    **A scope pattern for a very leaky transistor.**

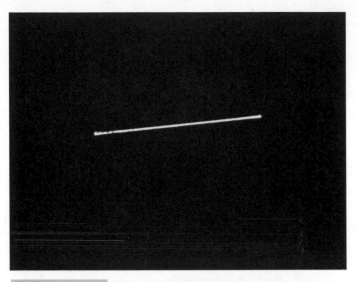

**FIGURE 1-39**   This scope pattern shows a circuit with a very low resistance.

# Electronic Power Supplies

All electronic devices must have some type of power supply or source voltage to operate. Most draw power from an ac power line and use rectifiers and filters to produce a dc voltage. Some equipment operates from batteries and the power supply is used to recharge the batteries. Most electronic circuits require a dc or direct current in order to operate.

### HALF-WAVE POWER SUPPLY

The circuit drawing in Fig. 1-40 is of a half-wave rectifier power supply. Notice at the top right, the negative going part of the sine-wave is missing and only the positive-going part is now available. The bottom waveform portion is removed by the diode rectifier because it only lets current pass in one direction. The pulsating dc is 60 times per second and is now smoothed out with a filter capacitor.

### FULL-WAVE POWER SUPPLY

The full-wave power supply circuit shown in Fig. 1-41 lets both halves of the ac sine-wave be used, with an output ripple of 120 times per second, rather than 60 times, as with the half-wave power supply. The two diodes are connected so that one diode conducts on the other half of the cycle. Thus, the diodes are conducting on each half cycle. This 120-cycle ripple now must be smoothed out with a resistor or iron-core choke and two filter capacitors. The choke helps prevent sudden changes of current through it and a second electrolytic capacitor (C2) provides even more filtering.

## A BRIDGE-TYPE POWER SUPPLY

Bridge diode power supplies are used in many kinds of electronic equipment, such as TVs, video recorders, and stereo sound systems. The bridge circuit power supply is unique because it can produce a full-wave output without using a center-tapped transformer. The typical diamond-shaped diagram for this type power supply is shown in Fig. 1-42. You could think of the bridge-rectifier circuit as an electronic switching system. Think of the diode rectifiers as switching all of the positive ac pulses to the B+ line and all of the negative ac pulses to the B-line or to chassis ground.

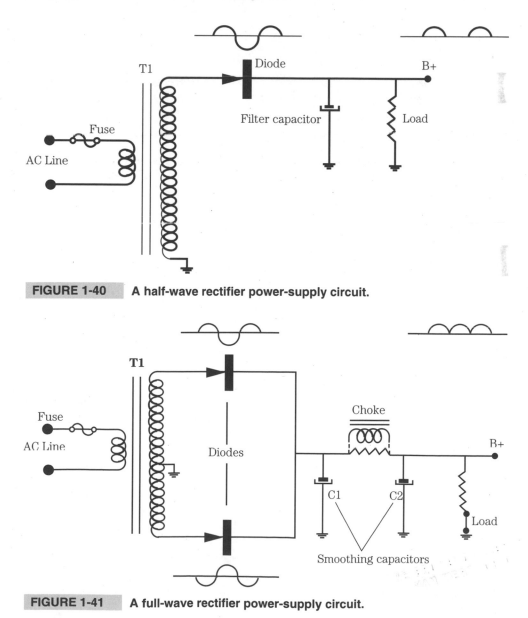

**FIGURE 1-40**    **A half-wave rectifier power-supply circuit.**

**FIGURE 1-41**    **A full-wave rectifier power-supply circuit.**

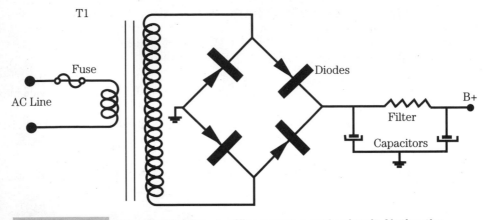

**FIGURE 1-42**    A typical bridge-rectifier power-supply circuit. Notice the diamond shape of the diodes' layout.

## THE VOLTAGE-DOUBLER POWER SUPPLY

A voltage-doubler power supply can have a transformer or it can be direct ac-line operated. The transformerless type is used in equipment that requires a higher dc voltage output and also to reduce the cost and weight of the device.

A basic transformerless doubler circuit is shown in Fig. 1-43. To see how it works, assume that the half-wave diode (X1) is connected to produce a positive voltage on the B+

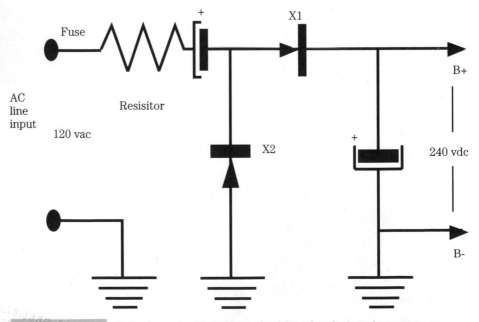

**FIGURE 1-43**    A power-supply voltage-doubler circuit that does not use a power transformer.

line of 120 volts. Diode X2 is then added to the circuit, but is connected in the opposite polarity. This will make diode X2 –120 volts, with respect to ground. This will "add," then produce a voltage of approximately 240 volts between the B+ and B– points. The problem with this power supply is that the B– is connected to the chassis of the device. This makes it a "hot" chassis, which will create a shock hazard. When you take the case or cover off of this type of equipment, always be cautious and you should plug the device into an isolation transformer.

Common power-supply problems are blown fuses, shorted diodes, burnt resistors, and open or shorted filter capacitors. Use your volt/ohm meter to check out the power-supply faults. A digital multimeter is being used in Fig. 1-44. You will find more power-supply information and circuits plus troubleshooting tips in other chapters of this book.

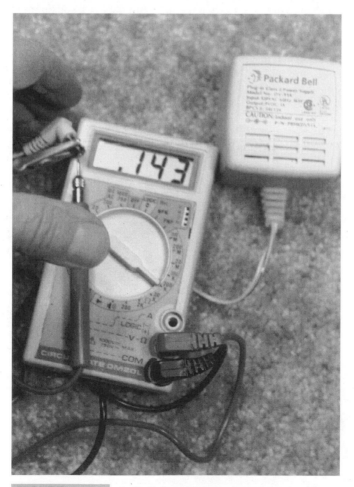

**FIGURE 1-44**    **A digital multimeter is being used to check a "block-type" plug-in power supply.**

# Electronic Circuit Soldering Techniques

When removing or replacing parts on a PC board, you will need a soldering iron (20 to 25 watts), rosin "flux" or solder with a rosin core. A solder wick, which is a flat-braided copper strips is a useful aid for soaking-up solder when removing a part from the PC board. Figure 1-45 illustrates how the solder wick is used to remove solder from a part on the PC board.

Figure 1-46 shows two types of soldering irons. The top one is a 25 watt and should be used for all PC board soldering. The larger 45-watt iron is used for soldering chassis grounds and large-wire connector lugs. Figure 1-47 is of a soldering gun and it only heats when the trigger switch is pulled on. These guns will heat up in about five seconds and usually are high wattage. They should not be used for soldering on PC boards because you can damage one very quickly. Many of these guns are rated a 100 to 150 watts. Figure 1-48 shows how a small iron is used to solder in the pins of an IC.

ICs can be directly soldered onto the PC board or they might have a socket mounted onto the PC board and the chip will plug into the socket. In Fig. 1-49, an IC is being removed from its socket. Very carefully pry up each end of the IC, a little each time, so as not to bend or damage the pins. When installing the IC, be sure that the pins are straight and are lined up with the socket pin holders. Also, be sure the key or notch is correctly lined up with that marked on the PC board. A chip put in backwards can be very costly. Be cautious and recheck position of the chip key.

**FIGURE 1-45**    **Solder wick being used to "suck-up" solder from a connection on a PC board.**

**FIGURE 1-46**    A 25-watt and a 45-watt soldering iron.

**FIGURE 1-47**    A fast-heating soldering gun.

**FIGURE 1-48**    A 25-watt iron being used to solder the pins of an IC.

**FIGURE 1-49**    An IC being removed from a plug-in socket located on a PC board.

## SURFACE-MOUNTED DEVICES AND THEIR SOLDERING TECHNIQUES

As surface-mounted devices (SMD) have evolved, the electronics industry have built SMD equivalents for most conventional electronic components. New electronic equipment contains SMD resistors, capacitors, diodes, transistors, and ICs. Even wire jumpers and 0-ohm resistors are used because they are more easily installed by automated assembly machines.

During assembly, the SMD unit is lightly glued to the circuit board with the metallic contacts lying on the copper path, where a circuit connection is to be made. Wave soldering then is used to join all SMDs electrically and mechanically to the board.

**Some SMD basics**  On most circuit diagram, an SMD device has an *M* following its part number. The *M* represents for (metal-electrode face bonding), which is the process used in producing chips.

Surface-mount components are available in various sizes and configurations, starting with large microprocessors, all the way down to single diode packages. Even single diodes and resistors are available in different sizes.

**Surface-mounted resistors**  A typical SMD resistor consists of a ceramic base with a film of resistive material on one surface. Refer to Fig. 1-50. Two electrodes are on the ends of the base, which is in contact with the resistance film. The contacts are used in making a solder connection to a PC board. The resistance of the device is determined by the amount of film material.

SMD resistors are typically in the ¼- to ⅛-watt range. The regular color code is not used on SMD resistors. Three numbers are usually printed on the film and give the same information as the color code. The first two numbers represent the first two significant numbers of its value. The third number represents the number of zeros.

**Surface-mounted capacitors**  Chip capacitors are fabricated with layers of resistance film, separated by layers of a ceramic base material, which is the dielectric. Notice Fig. 1-

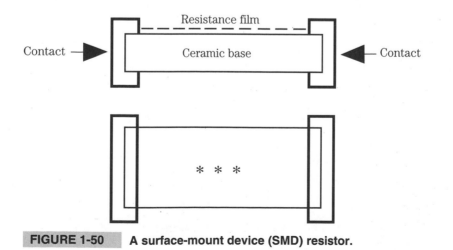

**FIGURE 1-50**    A surface-mount device (SMD) resistor.

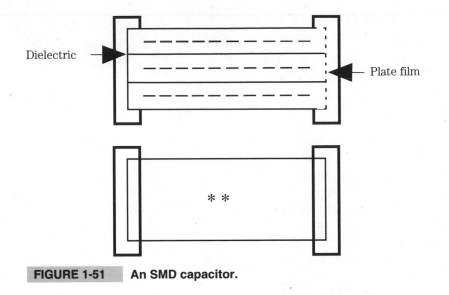

**FIGURE 1-51** An SMD capacitor.

51. The chip capacitor is very similar in appearance to the resistor. The body generally has a two-digit or two-letter code to show the capacitance of the device.

**Surface-mounted diodes and transistors** The SMD equivalent for solid-state devices are conventional silicon technology in new housings, again allowing for easier automated assembly. Refer to Fig. 1-52. The package for a diode is called an *SMC (single-melt component)*. The diode is marked on one end with a band to denote the cathode of the device.

The transistors are in packaging that corresponds to their purpose. The low-power device is in a SOT-23 (small-outline transistor) package. The transistors that function in heat-generating capacities are in a SOT-89 package that features a heatsink. The same packages are also used for FET and MOSFET devices.

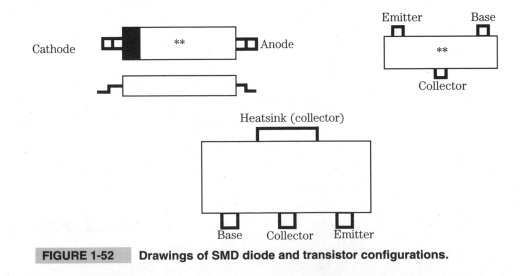

**FIGURE 1-52** Drawings of SMD diode and transistor configurations.

**Integrated circuits**    The SMD integrated circuit, like the diode and transistor, is conventional technology repackaged for automated insertion, as well as miniaturization of the circuit boards.

The SOIC (small-outline IC) is similar to the standard DIP packaging, except that the legs are designed for surface-mount soldering. Note layout of SMD IC in Fig. 1-53.

**SMD-soldering techniques**    Soldering of and/or replacement of an SMD is different from a standard component in two ways. First, the reduced size of SMD components and circuit-board paths increase the need for care when repairing this type equipment. Secondly, the tools required for repair are more specialized. Excessive heat can easily damage not only the SMD, but also the PC board paths. A controlled-heat soldering iron in the 20- to 25-watt range is a must. Small-diameter rosin-core solder is also needed. Solder wick is needed in different sizes and can be cut in short pieces. A bottle of flux should be used as an aid in heat transfer. Small-tipped tweezers and dental picks are useful in handling the SMD parts. A magnifier with a light source is very useful for close-up inspections. And a grounded soldering iron and tip should be used along with an anti-static wrist band to prevent damage to static-sensitive SMD components.

**Removing SMD resistors or capacitors**    In most cases, a SMD device is not reusable once it has been removed from the PC board. You should be sure that the device is defective using troubleshooting techniques before removing a SMD.

Now refer to Fig. 1-54. Add extra solder to the contact points to cause even solder flow. Grasp the component body with tweezers and gently rock back and forth while heating the solder on both ends. Remove the heat while continuing to rock the SMD contacts. Once leads are loose from the foil, quickly twist the SMD to break the epoxy or glue that was holding the SMD to the PC board.

**SMD transistor removal**    Refer to Fig. 1-55. Add solder to all three terminals. Grasp the component body with tweezers or needlenose pliers. Heat terminal C and rock the body up

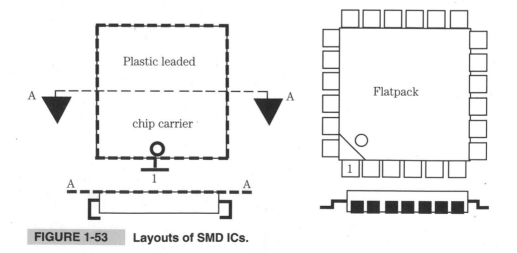

**FIGURE 1-53**    **Layouts of SMD ICs.**

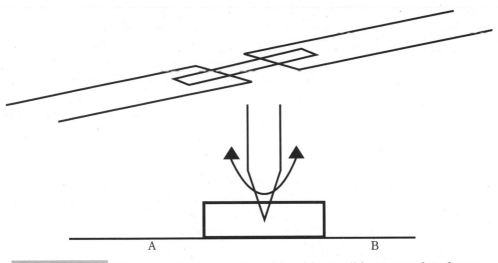

**FIGURE 1-54**    **For an SMD, always add extra solder to all contact points for an even solder flow.**

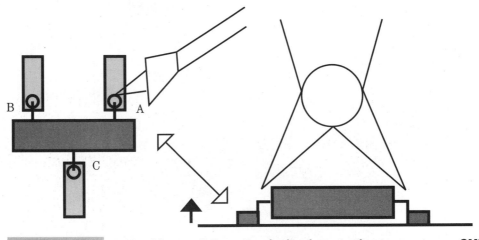

**FIGURE 1-55**    **Add solder to all three terminals when starting to remove an SMD component.**

until an open space exists between the terminal and pad. Now work on the other two terminals until loose.

**Removing SMD integrated circuits**    For IC removal, refer to Fig. 1-56. Apply solder liberally to all pins. Use a special soldering tip that will fit over the particular size of IC housing. This will allow all pins to heat up at the same time. Use a dental pick to lift the IC off as soon as the solder is molten.

**SMD parts replacement**    The replacement of any SMD follows a similar pattern. Be sure that the foil solder pads are free of any excess solder. Using short pieces of solder wick, clean

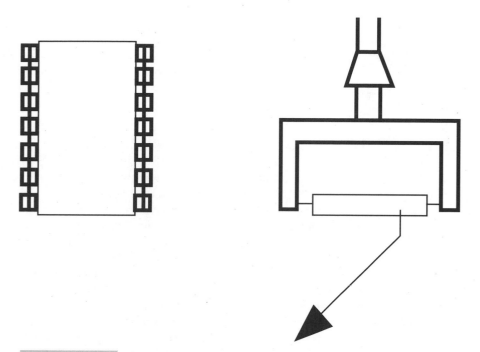

**FIGURE 1-56**    After all contacts of an SMD chip are heated equally, gontly pry up the device for removal.

**FIGURE 1-57**    A camcorder PC board with several surface-mounted components.

the pads until they are smooth. Spray with board cleaner, if necessary, to remove any residue of rosin. Position the device on the pads and hold them, as necessary, with picks or tweezers. Melt a small amount of solder on the tip of the iron. Then apply it to the lead. This will hold the component in place. Then, using the proper size of solder, attach all remaining legs. The photo in Fig. 1-57 shows a camcorder PC board with several SMDs.

# Electronic Test Meters (VOMs)

A volt/ohm test meter is a "must have" if you want to troubleshoot and repair any electronic equipment. These small, inexpensive meters can have an analog meter, which have a needle pointer that swings across the meter scale face plate, or a digital readout, which have the direct number readings on an LCD screen. Figure 1-58 shows some inexpensive digital read-out volt/ohm meters. You can find these meters at electronic parts supply stores, Radio Shack, Wal-Mart, and K-Mart stores. These test meters are called *multimeters* or *volt-ohm milliammeters (VOMs)*. These meters have pushbuttons or a switch to go from one function or rating to another.

You can use your voltmeter in the ac range to check voltage at wall sockets and where the ac line cord terminates in the equipment. You can check the ac power line voltage this way and see if the ac power is getting to the equipment power-supply section at the correct value. You can even locate open fuses and tripped circuit breakers with this ac voltage test. The dc range is used to check battery and charging voltage from any charger unit. Also check the dc voltage output from those small plug-in block power supplies. The dc range

**FIGURE 1-58**   Some inexpensive digital volt/ohmmeters that are very useful for electronic circuit repairs.

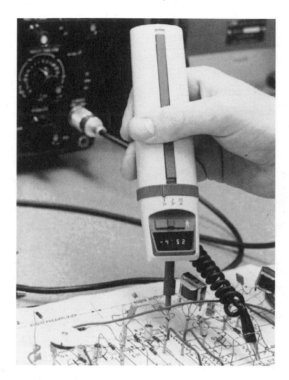

**FIGURE 1-59** A small, easy-to-use, battery-operated, portable volt/ohmmeter. Some cost $45 to $85.

is also used to check all of the other various dc voltage levels that are found in electronic equipment circuit boards. You can use the multimeter to look for low voltage, no voltage, or too high of a voltage.

A small, easy to use, portable, battery-operated volt-ohm meter is shown Fig. 1-59.

# Tools for Electronic Circuit Repairs

Now for some information on some common small tools that are very useful for electrical and electronic circuit repairs.

Diagonal cutters, sometimes called *side cutters* or *dikes*, are used to cut wires and component leads. They are also useful for stripping insulation from wires that are to be connected or spliced together. You should have two sizes of diagonal cutters (4" and 6") and long-nose pliers. The long-nose (needle-nose) pliers are used to insert parts, position lead wires and shape wires for connections. The common "gas" and utility "slip joint" adjustable pliers are also very useful. Figure 1-60 shows some of these basic electronic tools needed for repair work.

The following is a list of basic tools you should find useful for electronic repairs:

- Long-nose pliers.
- Diagonal cutters.
- Needle-nose pliers.
- Long-nose pliers with side cutters.

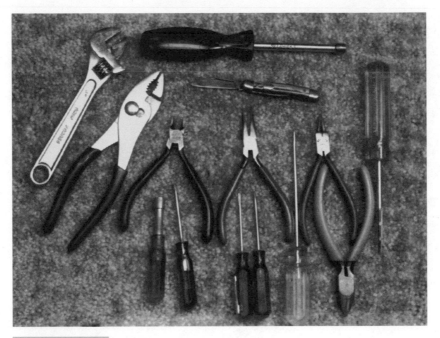

**FIGURE 1-60**   **Some basic tools you will need for electronic equipment repairs.**

- ■ Utility pliers.
- ■ Seizers, for holding and soldering small parts.
- ■ Electrician's knife.
- ■ Adjustable wrench (crescent).
- ■ Various screwdriver sizes and tips.
- ■ Nut drivers (spinners).
- ■ A small set of jeweler screwdrivers.
- ■ 20-watt and 45-watt soldering irons.

# Some Service Repair Tips

When working on your electronic equipment, it is very helpful to have service information and diagrams to reference. Some new equipment has information packed in the box or you can write to the manufacturer for this information. Also, books and schematic folders are available for the various models of TVs, VCRs, camcorders, etc. You can usually find these books and folders at electronic parts stores, such as Radio Shack, Allied Radio, and MCM Electronics. You can also order from TAB/McGraw-Hill Electronics Book Club.

Before you start working on your equipment with a problem you might want to make some notes and review the problem(s):

- ■ Notice when the problem occurs.
- ■ Is the device cold or hot when the problem occurs?

- How does it perform or not perform?
- How often does it occur? Is it intermittent?
- Have you had the electronic equipment repaired for the same symptoms before?
- Does it have to operate a long or short time before the trouble appears?

Thus, as you see from this list, you need to note any type hint or clue to solve these mysterious electronic problems. It helps if you are a good detective.

You will find that with most electronic equipment, such as CD players, video recorders (VCRs), camcorders, cassette players, or telephone answering machines, the problem is generally mechanical and not electrical. All you need for these repairs is a set of common tools, a cleaner/degreaser solvent, lubricating oil or grease, some alcohol for cleaning, and then just use your common sense for repairs. Some of these simple-to-repair problems are:

- For a CD player, check and clean the lens because it might be dirty.
- Also, for a CD player, check the lubrication. Check for oily slide drawer belts and dirt on the sled tracks or gears. A defective or partially shorted spindle or sled motor.
- For a VCR, check for broken or loose belts or belts that need to be cleaned.
- Also, for a VCR, clean the heads, the tape travel tracks, and rubber idler wheels.
- For any kind of video or audio recorder, look for a defective cassette tape cartridge, broken or tangled tape, tape wrapped around the capstan, and jammed-up parts.
- For all VCRs, audio tape recorders, and TVs, check for blown fuses, loose plugs and connections, and power-supply problems.

## INTERMITTENT TEMPERATURE PROBLEMS

Intermittent electronic problems are generally the toughest to pin down. Many of these faults show up after the equipment has warmed up. One trick you can try is using heat or a coolant spray (freon) to various small areas of the circuit board. A hair dryer is used in Fig. 1-61 to isolate a heat-sensitive component. This might take a little time, but you can solve the problem. The most common components to breakdown from heat or cold changes are ICs, transistors, diodes, and electrolytic capacitors. Also poor solder connections and PC board cracks can be located this way. And do not overlook small transformers, choke coil windings, and their connections.

## NOISY ICS OR TRANSISTORS

Often, the noisy transistor or IC can be located in the input and output sound stages TVs, CD players, and cassette system audio circuits. The hissing or frying noise that occurs with low audio levels can indicate a noisy solid-state component failure. Lower the volume level and listen for the frying noise. If the noise is still present, you know that the defective component is between the volume control and speaker.

You can try isolating the noisy component by grounding the input terminal of the power-output IC or transistor with a 10-ohm resistor to ground. With other transistor stages you can ground the base with a 10-ohm resistor, as shown in Fig. 1-62. If the noise becomes lower or disappears, you know that the defective component is before this stage. If the noise is still present, replace the transistor or IC in this stage.

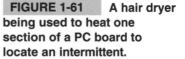

**FIGURE 1-61    A hair dryer being used to heat one section of a PC board to locate an intermittent.**

Sometimes spraying the suspected transistor or IC with a coolant spray will make the noise louder. Other times, the noise will disappear. At other times, again applying heat with a hair dryer on a suspected transistor or IC will make the noise reappear after applying another shot of coolant spray. Do not overlook the small ceramic bypass capacitors that can create noise when B+ voltage is on one side of this component. Replace the noisy component with a good part and then reheat or cool this same area again for a confirmation.

When the noise disappears with the volume control turned down, the noisy component will be ahead of the volume-control circuit. This transistor grounding technique can be used in other amplifier stages by jumping a 10-ohm resistor from the base to the emitter of the suspected transistor; if the noise stops, then the transistor is faulty. In a stereo audio amplifier system, start at the preamp input transistor and proceed through the circuit. If the noise is present after grounding out the first preamp signal, then the second preamp transistor must be noisy.

Usually, the noisy condition occurs in only one stereo channel. If both channels are noisy, suspect the stereo IC power output. The noise might disappear for several days, then reappear again. Replace the power output IC if a loud frying or hissing noise is present at all times. A poor internal transistor or IC junction is generally the cause of this type of noise.

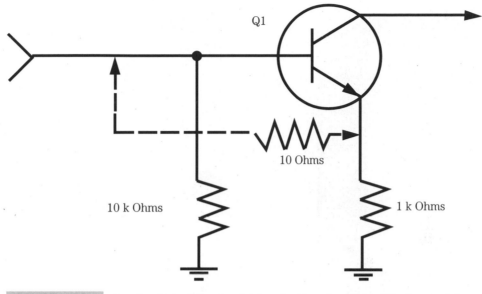

Q1

10 Ohms

10 k Ohms

1 k Ohms

**FIGURE 1-62**    A noisy transistor can be located by shorting a 10-ohm resistor between the base and emitter connections.

**FIGURE 1-63**    Do not start changing parts at random, also called "shotgunning."

Try to determine which part is faulty before replacing it, if at all possible. You don't want to start changing parts at random, also called *shotgunning*, like the fellow in Fig. 1-63, to solve a circuit problem.

You now know what components make up various electronic devices and how they work. You can now go onto the chapters of interest and solve the problems that occur in your equipment.

# RADIO/AUDIO/STEREO/SPEAKERS/
# MUSIC SYSTEMS AND CASSETTE
# PLAYER OPERATIONS

## CONTENTS AT A GLANCE

# Broadcast Radio Transmitter Operation

For you to become familiar with AM/FM radio reception, start by reviewing how the FM radio signal is developed and transmitted. An FM stereo signal must be compatible with monophonic FM radios, but they must also simultaneously carry other information, such as SCA background music, paging, and much more.

The two basic components needed for any stereo radio system are the right (R) and left (L) audio channel information. Refer to the basic stereo FM transmitter block diagram in Fig. 2-1. These left and right audio signals are matrixed, resulting in sum information (L+R) and difference information (L-R). Matrix is something within which something else originates or develops. To obtain sum information (L+R), +R was added to L; to obtain the difference information (L-R), a negative -R of the same magnitude as the +R (only 180 degrees out of phase) is added to L. Thus, L-R, the difference signal, was created. The composite L+R and L-R information is now used as FM modulating components in this system. Normally, the L+R information could immediately FM modulate the carrier. However, to be certain that the L+R information is in the same phase relationship to the L-R information, as they were when they came from the matrix when the FM modulated the carrier, it is necessary to insert a delay network in the L+R channel. The delay system is needed to shift the phase of the L+R modulating component in such a manner that it will be in phase with the L-R upper and lower 38-kHz sidebands when they also FM modulate the carrier.

In the FM stereo system of transmission, it is necessary that the L-R information AM modulate a subcarrier. To create this subcarrier, a very stable crystal oscillator produces a 19-kHz signal. The 19-kHz signal is doubled to obtain a 38-kHz subcarrier that is then AM

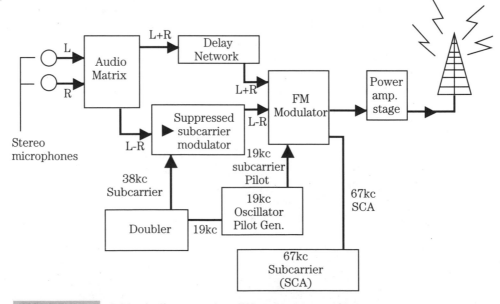

**FIGURE 2-1**    **A block diagram of an FM stereo transmitter.**

**FIGURE 2-2**    **The Bose AM/FM Wave Radio.** Courtesy of Bose Corp.

modulated by the L–R information. The 19-kHz signal is also used as a pilot signal or synchronization signal and it also FM modulates the carrier. Because all of the necessary signal information in the subcarrier system is contained in the upper and lower L–R 38-kHz sidebands of the AM-modulating envelope, the 38-kHz subcarrier need not FM modulate the carrier. Thus, the 38-kHz carrier is suppressed and only the remaining upper and lower L–R 38-kHz sidebands are used to FM modulate the radio carrier.

The FM broadcast system now has three carrier-modulating components: L+R audio information, two L–R upper and lower 38-kHz sidebands, and the 19-kHz pilot signal. As stated previously, it is necessary that these FM radio systems be compatible with facsimile or SCA. So, another modulating component, a 67-kHz subcarrier for SCA, needs to be added.

# FM/AM Radio Receiver Operation

The Bose Wave Radio, shown in Fig. 2-2, delivers sound quality for its small size that can't be compared to conventional radios or to ordinary stereo systems.

Linking a special configuration of Bose's unique waveguide technology and the Acoustic Wave Music System to a top-quality radio receiver, the Wave Radio generates sound far more spectacular than its compact size or the sum of its component parts would indicate. Despite its small size, the Wave Radio provides full, rich sound to fill most size home listening rooms. This remarkable audio breakthrough in sound quality comes from the 34-inch single-ended waveguide inside the unit. More on the Bose waveguide speakers later in this chapter.

All functions on the Bose Wave Radio can be regulated by a credit card-sized remote-control unit included with the radio. The Wave Radio features AM and FM stereo radio and a dual alarm clock modes. It offers 12 radio presets, mute, scan and automatic sleep features, as well as battery back up, in case of a power failure. You can set the Wave Radio so that you can fall asleep to one station and wake up to another. The volume will raise gradually to a volume level that you set.

## RADIO CIRCUIT OPERATION

Now see how the various circuits of a radio receiver operate and the problems that can occur. Refer to the block diagram in Fig. 2-3 as various sections are covered.

**The RF tuner section** The radio RF tuner selects the station you want to hear and also rejects any unwanted or undesirable radio signals or interference that is present at the antenna. The mixer stage is used to mix the RF station signal with the receiver's oscillator to produce the IF frequency. An AGC voltage is applied to the RF stage to reduce the amplification of this stage when a strong radio signal is being received. This AGC control voltage is developed at the detector and is usually used to control the amplification of stages in the IF and RF circuits of the receiver.

**Automatic frequency control (AFC) circuitry** With any high-frequency oscillators, stability is very important feature and these circuits require some type of AFC control to compensate for oscillator frequency shift. This is accomplished by taking a sample voltage from the ratio detector and feeding it via a *varicap*, a voltage-controlled variable capacitor, to the oscillator stage. The varicap is connected across the oscillator tuned circuit

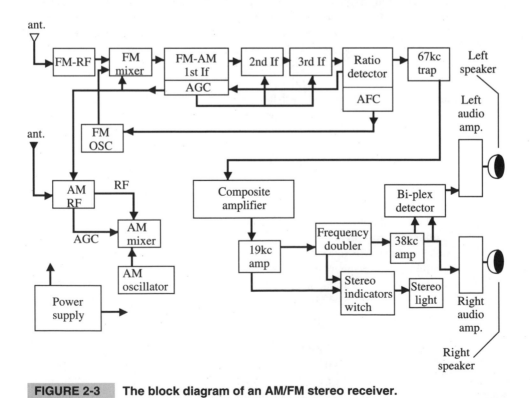

**FIGURE 2-3**   **The block diagram of an AM/FM stereo receiver.**

(a-d) acts as a frequency-controlling device. If the oscillator should drift, a ratio detector unbalance occurs and a dc voltage is fed back to the varicap so that its changing capacitance will automatically adjust the oscillator frequency. Thus, it has an automatic oscillator frequency control that eliminates drift and simplifies station tuning. Analog tuners will usually have an on/off AFC switch. When tuning in a station, turn off the AFC switch to disable the AFC control to more accurately tune in the station. The newer receiver tuners are digitally logic-IC controlled and do not have an AFC switch.

**Intermediate frequency (IF) amplifiers**   The FM IF frequency is usually 10.7 MHz and the IF frequency for the AM section is 455 kHz. The IFs in a receiver are used to amplify the RF signal and, with the addition of traps, make the receiver much more selective. The gain of the IF amplifiers is controlled by an AGC control voltage. The better receivers will usually have four stages of IF amplification. The processed signal is then fed to the FM ratio detector.

**Ratio detector AND composite amplifier**   The 10.7-MHz amplified output signal from the last IF stage is fed to the ratio detector. The ratio detector is a standard FM circuit that consists of diodes or a special detector chip. Assuming that the FM station you are tuned to is transmitting in stereo and with an SCA program, the composite output signals from the ratio detector will be:

- A 67-kHz SCA signal.
- A 19 kHz pilot signal.
- A L+R audio voltage signal.
- Upper and lower 38-kHz sidebands.

The composite signal goes to the input of a 67-kHz trap. If the FM station you are listening to is also sending out a 67-kHz SCA signal, it cannot be allowed to enter the detector or the audio will be very distorted.

**Composite amplifier function**   With the 67-kHz SCA information trapped out, it is now necessary to amplify the remaining parts of the composite FM detected signal. The composite amplifier has a gain of nine or more times. The output of this composite amplifier is fed to two channels. The L+R audio voltage and the 38-kHz L–R upper and lower sidebands are fed directly into the biplex detector and are then recombined with the developed 38-kHz subcarrier, as well as simultaneous detection into L and R audio voltages. The 19-kHz signal is usually taken off of a transformer and fed to the 19-kHz pilot amplifier.

Other circuits in a stereo FM receiver consist of a 19-kHz pilot signal amplifier, 19-kHz doubler, 38-kHz amplifier, and a circuit to indicate when you are receiving a stereo radio broadcast. This is called the *stereo indicator switch circuit*.

**Biplex detector operation**   Some receivers use a bilateral transistor in the biplex detector circuit to accomplish stereo signal separation.

For biplex detector operation, the (L+R) audio signal appears at the "L" and "R" output circuits in equal amplitude of the same polarity. With only a few turns in the 38-kHz transformer secondary winding, there is only a low-resistance path for the (L+R) signal. The

(L–R) 38-kHz sidebands are demodulated by the action of a transistor into two equal amplitudes, but with opposite polarity (L–R) regular audio signals in the same L and R output circuits. The biplex solid-state circuit thus acts to reinsert the 38-kHz contiguous wave (CW), which is a subcarrier into the (L–R) 38-kHz sidebands. At the same time, it demodulates this signal into the (L–R) audio signal and also provides the matrixing of the two sets of audio signals.

The demodulation efficiency of the multiplex "average-type" detectors is about 30 percent. The demodulation efficiency of the biplex detector circuit is near 60 percent. Furthermore, the L and R channel separation is improved to better than 6 dB at the higher audio frequencies between 8 kHz and 15 kHz. The biplex circuit is designed to provide about 25 dB of separation between the L and R channel signals at 1000 Hz.

One of the most desirable features of the biplex detector is that when tuning across the dial, both stereo and non-stereo (monophonic) stations are received at approximately the same volume level. During monophonic FM program transmissions, the 19-kHz pilot signal is not transmitted. If the 38-kHz switching signal is not applied to a switching transistor, it will remain turned off. In this case, the L+R audio signal will be divided between the two channels and fed to both the left and right audio amplifier channels.

The two stereo audio amplifier stages boost the signal level high enough to drive loudspeakers. They can be two or more speakers for each channel. The stereo amplifier stages will also have tone, loudness (volume), and balance adjustment circuits and controls for you to adjust to various room arrangements and to your listening preference.

**The Dolby recording technique**   First, see how an ordinary standard audio recording is produced.

***Making a standard audio recording***   Figure 2-4 illustrates how music consists of different loudnesses, separated by intervals of silence.

Loud and soft sounds are shown here as long and short lines. The music represented by this drawing starts loud and gradually becomes very soft and quiet.

Figure 2-5 represents noise. Any recording tape, even of the highest quality, makes a constant hissing noise when played. At very slow speeds and narrow track widths (used in cassette players), tape noise is much more noticeable than with a professional tape recording (although some noise is on these recordings, also).

Figure 2-6 depicts both noise and music on a tape recording. When a tape recording is played, the noise of the tape conceals the quietest musical sounds and fills the silence when

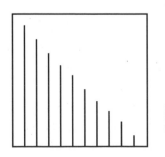

**FIGURE 2-4**   **The music, represented by this drawing, starts loud and gradually becomes very quiet.**

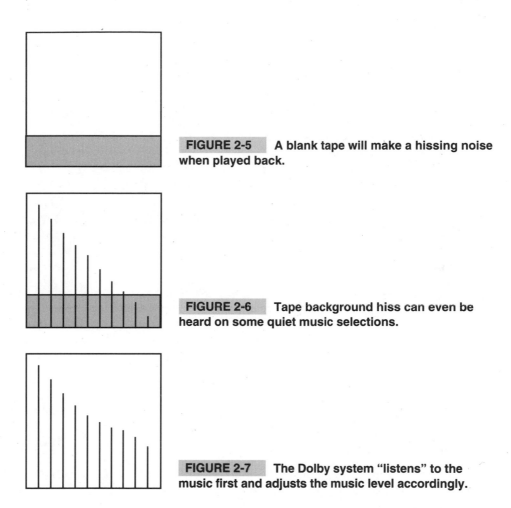

**FIGURE 2-5**    A blank tape will make a hissing noise when played back.

**FIGURE 2-6**    Tape background hiss can even be heard on some quiet music selections.

**FIGURE 2-7**    The Dolby system "listens" to the music first and adjusts the music level accordingly.

no sound should be heard. Only when the music is loud will the noise be masked and usually not heard.

However, tape noise is so much different from musical sounds that it sometimes can be heard even at these times.

***How a Dolby recording is produced***    Let's now see how the Dolby recording is made and what happens during tape playback.

**The Dolby system "first" listens**    Before the tape recording is made, as shown in Fig. 2-7, the Dolby system "listens" to the music to find the places where a listener might later be able to hear the noise of the tape surface. This happens mainly where the quietest parts of the music are recorded. When it finds such a place, the Dolby system automatically increases the volume being recorded so that the music is recorded louder than it would be normally.

Figure 2-8 gives you an indication of what the Dolby system is doing during recordings. In a Dolby system, recording the parts of the music that have been made louder, stand out clearly from the noise.

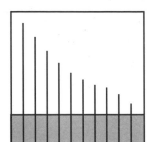

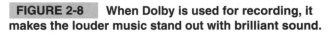

**FIGURE 2-8**    When Dolby is used for recording, it makes the louder music stand out with brilliant sound.

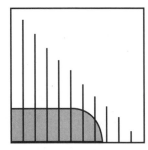

**FIGURE 2-9**    When a Dolby recorded tape is played back on a Dolby machine, the loudness is automatically reduced in all places that it was increased before.

As a result, the Dolby system recordings sound brilliant and usually clearer—even when played back without the special Dolby system circuit.

What the Dolby system does during playback is illustrated in Fig. 2-9.

When the tapes are played on a high-fidelity (hi-fi) tape recorder equipped with the Dolby system circuitry, the loudness is automatically reduced in all of the places at which it was increased before recording. This restores the music to its original loudness once again.

At the same time, the noise that has been mixed with the music is reduced in loudness by the same amount, which is usually enough to make it inaudible.

# Tips for Making Your Audio Sound Better

Of course, the placement of your stereo speakers is a very personal matter, depending mainly on the arrangement and layout of your listening room, speaker positions, and the way you listen to music. Where you place your speakers does make a difference in how your system will sound. Before settling on a final arrangement, try several arrangements.

Bass response is very dependent on speaker location. For maximum bass, place the speakers in the corners of your room. Placing the speakers directly on the floor will produce an even stronger bass response. If the bass sounds boomy and exaggerated, move the speakers away from the corners slightly, pull them out from the wall, or slightly raise them up off the floor.

## POSITIONING YOUR STEREO SPEAKERS

Stereo speakers should be placed from 6 to 8 feet apart. Putting them too close together reduces the stereo effect, but placing them too far apart reduces bass response and creates a "hole effect" in the middle of your room. Generally, most speakers have a tweeter dispersion angle of close to 60 degrees. For this reason, your listening position should be in the overlap zone, so you want to angle the speakers toward you for better stereo sound.

# FM Radio Antennas

Usually, the built-in antennas in most receivers are adequate for good reception. However, if you are having reception problems, try the following hints.

For better FM reception, you can build the folded dipole shown in Fig. 2-10. Just splice together 300-ohm TV twin-lead, as shown. Apply a small amount of solder and heat to the

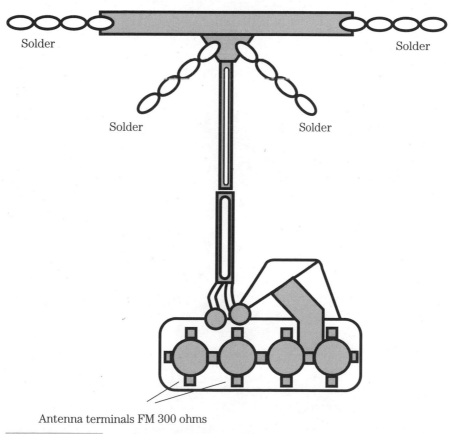

4 ft 8 in (142cm)

Solder    Solder

Solder    Solder

Antenna terminals FM 300 ohms

**FIGURE 2-10**    **An FM dipole antenna that you can make.**

twisted ends until the solder flows over each wire strand. Attach the lead-in to the 300-ohm terminals on back of the receiver. The antenna can be stapled or tied to the back of the receiver or placed on a wall. Turn or move the antenna around for the best reception.

You can use a set of TV rabbit ears, or buy an FM antenna. Some deluxe antennas feature electronic "tuning" for a more directional station reception.

An outside VHF/UHF/TV antenna will also work well for your receiver. A "splitter" will let you connect a TV and FM receiver to the same antenna. If you live in the country side, a specially designed FM antenna can receive an FM station from greater than 100 miles away.

# Some Receiver Trouble Checks and Tips

Now go over some radio receiver problems.

### RECEIVER WILL NOT OPERATE AT ALL

If your receiver is completely dead, check the power supply with a dc voltmeter for a presence of B+ voltage. If there is no B+ voltage, check for an open fuse (Fig. 2-11). If your radio has a built-in cassette tape deck and/or CD player and they are working ok, then the

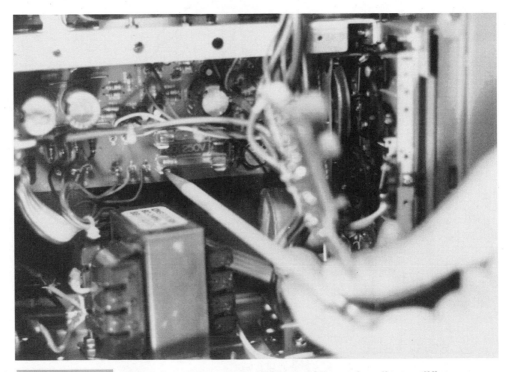

**FIGURE 2-11**   Check for a blown fuse if the receiver and audio amplifiers are dead.

power supply should be working and the problem is in the radio RF tuner, IF stages, or de-tector/multiplex circuit stages. To repair these stages, you need a voltmeter (VTVM), tran-sistor/diode checker, signal tracer, and oscilloscope. These repairs require a professional electronics shop or technician.

If the receiver or cassette/CD player has no audio, then the problem would be in the au-dio power amplifiers or speakers if the power supply checks out OK. Amplifier problems could be caused by poor solder connections, cable plug-in sockets, defective ICs or tran-sistors, open coupling capacitors, or burned (open) resistors and coils. An audio signal tracer can be used to isolate a loss of audio in these amplifiers. When you are signal trac-ing in either the RF, IF, or audio stages, the dead stage will become apparent when a sig-nal is found at the input of a stage, but not at the output of that stage.

Another quick check is to place the tracer probe at the speaker-output coupling ca-pacitors. The same signal should appear at both ends of the capacitor if it is good (or shorted), but an open capacitor will have an input signal, but no signal at its output con-nector. Open electrolytic coupling capacitors between the output stages and the speaker are fairly common. So, if the left or right audio channels are dead, you should check this out first.

If no signal is found at either end of the speaker coupling capacitor of the dead channel, move the probe to the driver stage output and then to the input. A signal at the output but not at the input proves that the stage is defective.

## INTERMITTENT RECEIVER PROBLEMS

Signal tracing is effective if the intermittent condition can be induced. To speed up the break down, you can use a heat gun (hair dryer) or some cooling spray to make the inter-mittent condition start or stop. After you find that a thermal condition triggers the fault, the heating and cooling should be applied to a small circuit area until the trouble can be pin-pointed to one component.

Capacitors are a common cause of intermittents. They can become intermittently leaky, shorted, or open. A leaky capacitor can change the bias on a transistor or IC and cause it (and other components) to fail. Some intermittent problems will change the B+ voltage, so closely check this to determine if voltage change might be the cause or the effect.

Other receiver intermittents are caused by various controls and switches that need clean-ing. Check the front-panel controls and switches. Notice the push-button switches in Fig. 2-12 and, while operating them, listen for any intermittents. These controls and switches can be cleaned with a special spray contact cleaner. If spraying with a cleaner does not cor-rect the intermittent problem, the control or switches will have to be replaced.

If the station tuning dial will not move the pointer, the cord is probably broken. If it is broken, you can replace the cord by restringing it. The tuning cord is shown in Fig. 2-13. If the cord is slipping, you can apply some anti-slip liquid or stick rosin compound.

## SOME RECEIVER SERVICE DON'TS

When working on solid-state (transistors and ICs) receivers, key voltage and resistance checks can usually be used to find the fault. However, before you start probing around,

**FIGURE 2-12**    Clean the selector switches for intermittent or noisy operation.

**FIGURE 2-13**    If the station selector will not turn, check for a broken or slipping string or belt. Replace broken string and apply anti-slip stick or liquid to the string.

taking measurements, and replacing components, you should look over the following rules.

■ Don't probe around in a receiver plugged into the ac socket. A short from base to collector will usually destroy a transistor or IC. Many stages are direct coupled, thus lots of components can be damaged. Always turn off power to the receiver before connecting or disconnecting the test leads.

■ Don't change components with the power applied to the receiver. Always turn off power to the receiver, except when taking voltage readings and then be very careful.

■ Don't use test instruments that are not well isolated from the ac line when making measurements on equipment connected to the same power source (even if the equipment is turned off). This prevents cross grounds. Check all test instruments and use an isolation transformer on the receiver under test.

■ Don't solder or unsolder transistor or IC leads without using some type of heatsink clip. This prevents damage to heat-sensitive solid-state components. Long-nose or needle-nose pliers make a good heatsink for soldering.

■ Don't arc B+ voltage to ground because transient voltage spikes can ruin ICs and solid-state devices fast.

■ Don't short capacitors across another capacitor or circuit component for a test. This can also cause ICs and transistors to be damaged.

■ Don't forget that many solid-state stages are directly coupled. A fault in one stage can cause failure in another stage.

■ Don't use just any ohmmeter for resistance checks. The voltage at the test probes can exceed the current or voltage limits of the solid-state device under test. The lower resistance scales on 20,000 ohms-per-volt meters are usually safe for short- or open-circuit checks.

■ Don't forget to reverse the leads when making in-circuit resistance checks. The readings should be the same either way. A different reading usually means that a solid-state junction is affecting the reading. You are actually making a check across a junction in a transistor or IC.

■ Don't forget to use extra caution when checking, unsoldering, or inserting and resoldering MOSFET transistors or MOS ICs.

# Loudspeaker Concepts and Precautions

One of the most important components of a good audio system is the speaker and its enclosure. This section covers speaker operation, how they are connected, speaker enclosures, and tips on hooking up your speaker system.

## HOW SPEAKERS ARE CONNECTED

Figure 2-14 shows connections for a woofer, mid-range horn, high-frequency tweeter, and a crossover network for a typical speaker system. In this set up, the crossover coil (1-mH choke) is installed in parallel with the 1-kHz exponential horn after the series crossover capacitor. In other systems, you might find the horn connected in series with the tweeter. The crossover point is usually at the 1-kHz frequency point.

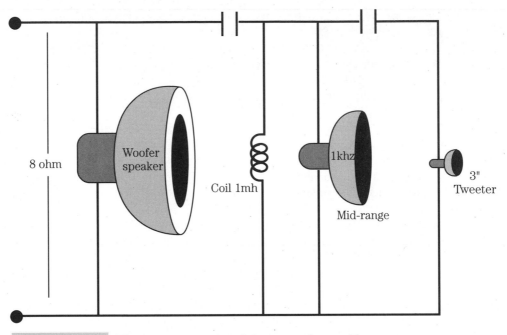

**FIGURE 2-14**    **The crossover network for a speaker and horn.**

The purpose of the crossover coil is to shunt the heavy bass frequencies around the 1-kHz horn and tweeter, thus affording added protection for the voice coils of the horn and tweeter. The crossover coil also serves to smooth the horn's acoustic crossover point and improves the speaker's sound reproduction.

A brief analysis of the speaker system (shown in Fig. 2-14) is as follows: Assume that the complete audio spectrum appears at the input of the 8-ohm speaker system. All frequencies will appear across the 12-inch woofer. The 12-inch woofer will reproduce audio frequencies from 30 Hz to 1 kHz. The crossover capacitor will block virtually all frequencies below 1 kHz and pass the audio frequencies above the 1 kHz (crossover) point. The 1-mH choke will act as a very low impedance to any frequencies below 1 kHz that might still be present after the blocking capacitor while acting as a high impedance to the frequencies above 1 kHz. Audio frequencies above 1 kHz will now be present across the 1-kHz horn. The acoustic audio output of the 1-kHz signal is essential flat, out to approximately 8 kHz. The capacitor blocks the frequencies below 8 kHz and passes the higher frequencies across to the 3-inch tweeter for a smooth acoustic output to approximately a frequency of 16 kHz.

If you are connecting new speaker to your audio system, be sure that they match for impedance (such as 8 or 16 ohms) and have proper power-handling capability for your power amplifier. It is possible to damage a speaker system—even if its power-handling capacity is the same as, or higher than, the power output rating of the amplifier to which it is connected. Damage can occur to the speaker system because almost all power amplifiers deliver more than their rated power output. This is especially true if the amplifier is operated at maximum, or very high volume settings while the tone controls are set at, or near, maximum boost. A safety margin should be allowed between the amplifier's rated output and the speaker system's rated handling capability if the amplifier is to be operated at very high volume levels. If distortion is noticed at high volume levels, it is recommended that the volume level be reduced because you might be reaching the safe operating limits of the amplifier or the speaker system.

## HOW TUNED-PORT SPEAKER SYSTEMS WORK

The speaker system shown in the (Fig. 2-15) drawing is of the tuned port type. The cross section view of this speaker enclosure is shown in (Fig. 2-16) and will be used for the following explanation. This enclosure has four openings in the front panel (one for each of the three speakers and one for the port), and the remaining panels are of solid construction.

A tuned-port speaker can be described as a tuned enclosure in which the air in the port will resonate with the air in the main area of the cabinet, at a given frequency. This frequency determines the effective low-frequency cutoff of the system (cabinet and enclosure combined). Below the selected frequency (30 Hz), the response drops very rapidly (approximately 24 db/octave).

This system could also be described as an acoustic phase inverter. That is, at some frequency, within its normal operating range, the air in the port is moving in an outward direction (to the front) while the speaker cone is also moving in an outward direction (to the front). These two movements would occur at the same time, and in phase.

Basic advantage of a tuned-port enclosure, as used in this speaker system, over a typical acoustic suspension (closed box) enclosure are:

■ Reduced low frequency distortion.
■ Increased efficiency. This requires less amplifier driving power for equal loudness level.

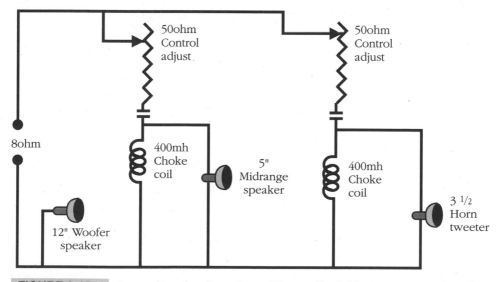

**FIGURE 2-15**    **A speaker circuit system with an adjustable crossover network.**

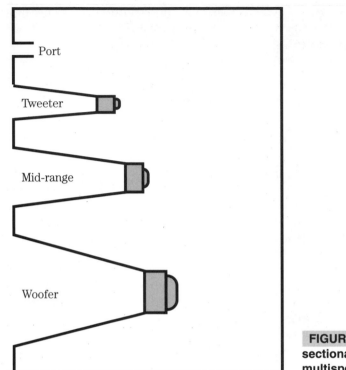

**FIGURE 2-16** A cross-sectional view of a typical multispeaker enclosure.

Sound level of signals radiated through the port (in the 30- to 70-Hz range) is comparable to the sound level radiated by the woofer (in the range of 70 Hz to 1 kHz). For the 15-inch woofer, a 3½-inch diameter port is required and it must "pulsate" air at a much higher velocity than a woofer with a diameter of 12 inches. Several factors must be considered to maintain the required port velocity.

■ The woofer uses a highly efficient magnetic structure, making it comparable to a powerful electric motor. This forces air in the port to move at high velocities—even though the air in the box is attempting to stop the motion of the speaker

■ The internal air pressure in a tuned-port enclosure is much higher than that in a conventional closed-box enclosure. The mechanical construction of a tuned port enclosure must be built better than for an air-suspension type.

■ Speakers (and other components) must be securely fastened to prevent air leaks. Leaks or loose components can result in losses, which cause a deterioration of performance. To increase effective cabinet volume and also to dampen internal resonances of the enclosure, acoustic padding is placed on the inside surfaces. These pads must not obstruct the port.

When servicing or connecting the amplifier systems, always be sure that the speakers are properly phased. If both speakers are phased in the same way, poor channel separation will result. Also, some loss in midfrequency response will probably be noticed. If the two input speaker signals are not 180 degrees out of phase, the bass frequencies will cancel, resulting in poor bass performance.

The size (gauge) of the speaker wires are also important for proper speaker sound reproduction. This will be very noticeable on high power amplifiers at high volume levels. A very small, 28-gauge speaker wire could distort and, in some cases, damage the amplifier output stage or even the speakers.

**The Bose Acoustic Wave speaker system**   Two of the chief differences between the technology used in the Bose Acoustic Wave stereo introduced in 1985 and that used in the Wave Radio today is the positioning of the loudspeaker and the length of the waveguide. The speaker in the Acoustic Wave is located so that one-third of the tube is behind the speaker and two-thirds are in front of it. In the Wave Radio, the speaker is nestled at one end of the waveguide, a less-efficient position, but the only one possible in a device this size. The Acoustic Wave houses 80 inches of waveguide; the Wave Radio cabinet enfolds 34 inches of ductwork. The longer waveguide in the Acoustic Wave system makes it possible to reach deeper into bass, to perhaps 40 Hz.

Another example of the Bose advanced systems is the all-in-one system format called "Lifestyle 20," which represents a quantum leap in both performance and convenience over other systems. A photo of the Bose "Lifestyle" system, including the jewel cube speakers is shown in Fig. 2-17.

**Bose Series III Music System**   Another great Bose audio center, shown in Fig. 2-18, is the Acoustic Wave Music System III. The system, measuring about 10 inches high by 18 inches wide and 6 inches deep, includes a full-featured CD player, AM/FM stereo tuner with 10 presets, and all the speakers, amplification, and equalization technology to fill a room with concert hall sound.

The newest version of this system provides even smoother audio performance and a slim remote that can operate the unit from 20 feet away. Other user-friendly features are color-coded, one-touch button operation, volume protection, and a continuous music option.

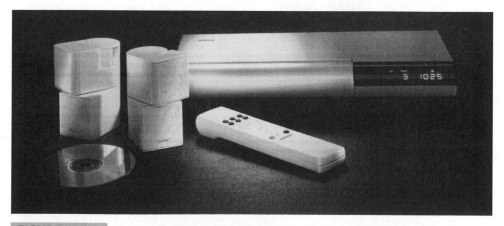

**FIGURE 2-17**    **Bose Lifestyle 20 music system and jewel cube speakers.**
Courtesy of Bose Corp.

**FIGURE 2-18**    **Bose Acoustic Wave music system.** Courtesy of Bose Corp.

**FIGURE 2-19**    **A cutaway view of the Bose waveguide chamber design.** Courtesy of Bose Corp.

Waveguide technology is based on controlled interaction of acoustical waves with a moving surface. This interaction occurs inside the precision waveguide—a mathematically formulated tube inside of which a loudspeaker is placed. The waveguide inside the Acoustic Wave Music System is nearly seven feet long and folded numerous times to fit inside the enclosure.

The cut-away view drawing in Fig. 2-19 shows the sound-channel configurations and the 36-inch-long waveguide inside a 14-inch case enclosure. This Bose acoustic wave system, which is about the size of a briefcase, has a tube length of 80 inches. The waveguide precisely matches the specifications of the speaker and skillfully controls the flow of air. This is how Bose is able to produce rich, full sound from unassuming small equipment.

**Bose Lifestyle 901 System**   Combining Bose's best loudspeaker with its most advanced systems technology, the Lifestyle 901 music system, shown in Fig. 2-20, is intended to come as close as of today to the sound of the original live performance.

The Lifestyle 901 system resulted from 12 years of physical acoustics and psychoacoustic research at the Massachusetts Institute of Technology. Many of the design improvements represented technological challenges for Bose engineers. The desire for high power handling and better efficiency produced two major achievements: the helical voice coil (HVC) driver and the Acoustic Matrix enclosure.

The HVC driver uses aluminum edgewound on an aluminum bobbin. The design allows significantly more windings on the bobbin without the air gaps caused by round-wire windings. The result is an efficient driver with high power handling.

Furthering efficiency and enhancing bass performance became the challenge of the Acoustic Matrix enclosure. By porting each of the nine drivers, air from the back of the cone could be used to increase efficiency and provide even deeper bass. Original designs produced the desired effect, but with undesired port noise. The final design is an injection-molded plastic enclosure that ports each of the drivers into a separate chamber, which, in turn, is ported through one of three reactive air columns. This sophisticated approach again required Bose engineers to design a manufacturing process from scratch.

The 901 speaker performance is optimized through integrated electronics, including amplification, signal processing, and active electronic equalization. Knowing the performance parameters of the 901 speaker allowed Bose engineers to match the ideal amplifier to achieve the renown room-filling sound of the speaker. Highly sophisticated system-protection circuitry ensures that the speaker is never over-driven and prevents interruption of the radio programs or music.

The system is controlled by the music center and integrated signal-processing provides deep, well-defined bass at all listening levels—even background levels—by compensating for your ear's decreased sensitivity at low volumes.

**FIGURE 2-20**   **Bose Lifestyle 901 music system.** Courtesy of Bose Corp.

**FIGURE 2-21**    **Bose Acoustimass 10 Home theater speaker system.**
Courtesy of Bose Corp.

**Bose Home Theater system**   The Bose 10 home theater speaker system is shown in Fig. 2-21. The Acoustimass 10 home theater speaker system includes five Bose signature double cube speaker arrays, a single Acoustimass module that can be hidden anywhere in the room, and unique, easy-to-use connectors. The system is compatible with all digital and analog surround sound electronic formats.

Bose technology allows a single unobtrusive Acoustimass module to provide pure low-frequency sound to the front and rear channels in the system. Virtually invisible cube speaker arrays produce consistent spectral and spatial perspective for front, right, center and rear channel sound.

If the Acoustimass 10-cube arrays are practically invisible, the system's low-frequency module can be also. It is small enough to be hidden anywhere in the room, but its deep bass performance is sure to be noticed. The module launches sound waves from three high-performance 5¼-inch drivers into a room in the form of a moving air mass, unlike conventional systems that rely on the vibration of a speaker cone. The result: pure sound, wider dynamic range, and virtually no audible distortion. A built-in protection circuitry guard's system components against excessive volume input levels.

# Cassette Players—Operation and Maintenance

The audio cassette players use a cartridge with two reels mounted inside a plastic holder. The tape is slightly more than ⅛-inch wide and is used for monaural and stereo audio recordings. The cassette tape is a thin plastic film coated with a layer of brown metallic dust (oxide). During recording, the oxide is given a detailed magnetic code. During playback, the tape passes over the record/play head with a head gap between two small magnets. The magnetic code on the tape changes the magnetic field at the head gap and the recording is decoded. When the head gap is clean, the tape undamaged and its speed is correct, you will hear the "live" sound intended by the musicians and recorded by the sound engineer.

A mono recording consists of two tracks, each 0.59 inches wide, separated by a guard band of 0.011 inches. In addition, a 0.032-inch guard band separates each pair of tracks, thus ensuring playback compatibility of stereo tape recordings. Total track width of each stereo pair, plus their guard band, is equal to one mono track width. Stereo prerecorded tapes will be reproduced in mono on a mono tape recorder unit. Left and right track signals will be combined by the playback head and be reproduced as a mono program. Recordings that are made on a mono unit will be reproduced only as mono sound—even on a stereo playback unit.

## GENERAL CASSETTE CARE

Cassette operation is very much the same as for the original reel-to-reel tape recorders, except that the cassette reels are smaller and enclosed within a molded plastic cassette housing. When being operated, the tape will be unwound from one reel supply, move past the tape heads and pressure pads, between the capstan and pinch roller, and finally on to the take-up reel. Movement of the tape by the drive system will stop when the tape is fully wound onto either reel in the cassette cartridge. A simplified drawing of the cassette recorder and tape path is shown in Fig. 2-22. The direction of the tape motion is determined by the function buttons. The cassette also has a window, through which you can estimate how much tape and playing time remains.

To prolong tape life, store the cassettes in a clean, cool, dry area in a closed container that will protect them from dust and moisture. Each cassette should be stored in its original container because this will help prevent dust and other materials from entering and causing possible tape damage. Avoid storing tapes after running at fast forward or rewind because this tends to create an unevenly wound tape.

Layers of tape will be compressed or loose, and wavy tape in addition to creating extra segments of tape by stretching it slightly its structure. All of this creates extra wow and flutter. For the same reason, it is good to play the cassette at least a few times a year. Do not store cassettes next to a heat source or stray magnetic fields. In warm summer climates, do not store in an auto or in direct sunlight.

## CASSETTE TAPE CIRCUIT OPERATION

A block diagram of a typical cassette tape unit is shown in Fig. 2-23. Notice the audio signal input and output jacks on the left side of the drawing.

**The Play mode**  With the cassette unit in the Play mode (Dolby noise circuit off), the audio signal moves from the Record/Play head via the Record/Play switch and into the play-

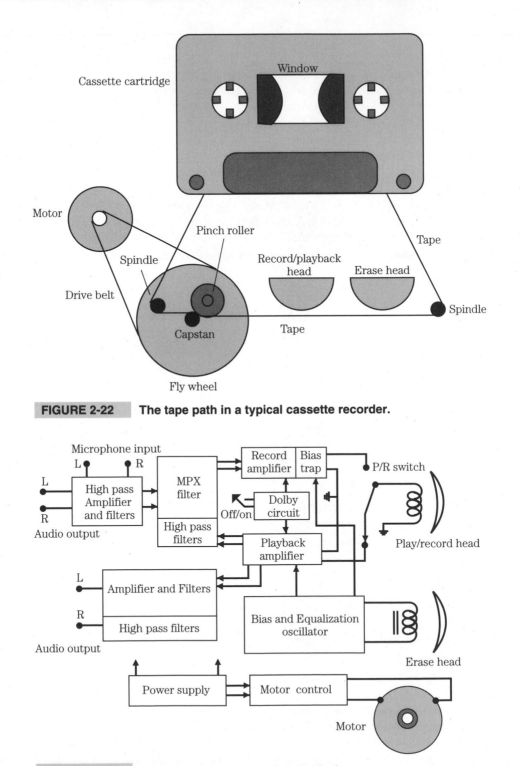

**FIGURE 2-22**    **The tape path in a typical cassette recorder.**

**FIGURE 2-23**    **A block diagram of a cassette deck.**

back amplifier block. A desired signal feedback then goes to the equalization block during playback operation. The audio processed signal then goes into more amplifier and filter block circuits and then is fed via jacks and cables to the audio power amplifier to the speakers of your audio system.

Switching the "Dolby noise-reduction circuits" ON, while in the Play position, will result in added operating circuits which control the dynamic processing characteristics and reduces the tape noise level. The audio signal is routed through the noise-reduction circuits and some high-pass filter stages. The high-pass filter attenuates the low and midrange frequencies.

**The Record mode**   During the Record mode, microphone audio is processed by an IC amplifier and other external audio by another set of input jacks and IC pre-amplifiers.

In the Record mode, when the Dolby noise-reduction circuits are switched on the audio goes to the high-pass filter stage. The high-pass filter functions in a similar way as in the Play mode, but is part of a positive feedback loop, instead of a negative feedback loop used in the Play mode. This will result in record circuit characteristics that are complementary to that of tape playback.

From the record amplifier, the audio signal goes through a bias trap. More on the bias oscillator later. This trap prevents the bias oscillator signal from getting back into the record amplifier or other circuits where the bias frequency could cause some undesired effects.

**Equalization circuits**   Equalization circuits are needed because of the different types of recording tape available. Some tape decks have equalization provisions for recording and playback for several types of tapes, such as ferric oxide, ferri chrome, and chromium dioxide. Your more-expensive tape decks will have switches or buttons on the front panel to adjust the unit for proper bias and equalization to match these various types of tapes. This tape-type equalization should not be confused with the normal record and playback equalization provided on all machines for proper reproduction.

**Tape player electronics**   Most modern cassette players now have all of the electronic components mounted on one PC board. These components will consist of capacitors, resistors, diodes, and ICs. The power supply might be found on this board or be located in another section of the audio system. If the tape unit is dead, then check out the power supply for correct voltages. Also, check any fuses in the power-supply section that might be blown.

On some tape recorders, you will find a automatic level control (ALC) circuitry located on the main board and it will have an adjustment marked (ALC Adj). An ALC circuit not working or not adjusted properly should be suspected when audio playback has distortion or changes in recording levels are being noticed. A frequency-compensating network is incorporated within the amplifier PC board circuitry, providing equalization required for proper record/playback response of the tape composition.

**Bias oscillator operation**   The bias oscillator circuitry serves two functions in a recorder. One is to supply erase current to the erase head while the second function is to supply record bias current to the play/record head. A pre-recorded tape must be cleanly erased to make another good tape recording on it. Bias current varies with the recording-level bias adjustments. The bias current is combined with the audio output signal from an IC amplifier, after which it is fed to the respective left and right windings in the play/record head. This bias current signal is required to make an magnetic audio tape recording. The bias oscillator current is usually generated from an IC on the PC board. Some recorders will have two bias control adjustments to establish the correct recording bias level and playback level.

**Cassette belt and rubber pulley drive systems**   You will usually find several belts and rubber drive wheels within any cassette tape mechanism. Most will have a motor drive belt to the capstan and flywheel assembly (Fig. 2-24). The drive belt is very small in some tape players. Some motor drive belts are only two or three inches long. The belts can be flat, round, or square in shape. Besides the motor drive belt, another belt runs from the flywheel to the take-up reel. You might find a fast-forward belt drive on some cassette players. Some of these belts are slim and not very thick, so they can stretch and cause erratic speed and wow. Clean each belt and drive wheel when you encounter speed problems with alcohol on a cloth. When the belts have been cleaned and you still have an erratic speed problem, then you need to replace the motor drive belt. A photo of a belt drive and flywheel is shown in Fig. 2-25. A typical cassette belt and drive arrangement is shown in Fig. 2-26.

Slow tape speed can be caused by a slick or dirty motor drive pulley. A dry capstan/flywheel bearing can cause the tape to run slow. A worn, stretched, or greasy belt can be the

**FIGURE 2-24**   **Cassette drive belts and gears.**

**FIGURE 2-25**    The flywheel and belt drive.

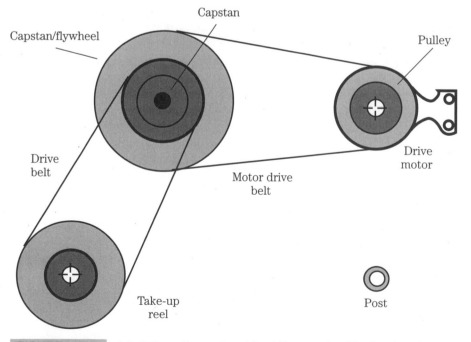

**FIGURE 2-26**    A belt from the motor drives the capstan/flywheel and the take-up reel assembly is belt driven from a small pulley on the capstan hub.

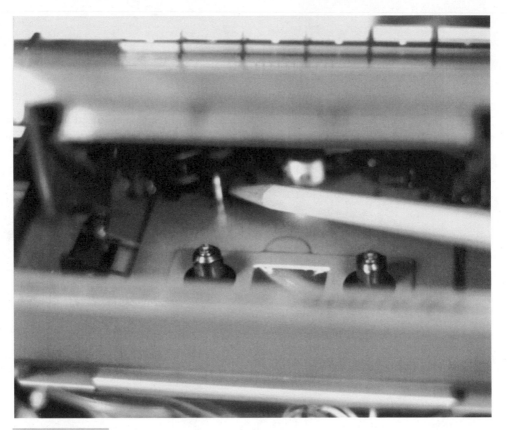

**FIGURE 2-27**    The tape can wrap around the capstan spindle. Remove the tape and clean the spindle with alcohol.

cause of slow speed, also. Clean any of the packed tape oxide from capstan, belts, drive wheels, pinch roller, or the play/record heads. Slower speeds can result if excessive tape is wrapped around the pinch roller and in between the rubber roller and bearing mount. Also, if the tape has broken, you might find it wound around the capstan spindle Fig. 2-27. Dig the tape out and clean the capstan with alcohol. Suspect a faulty drive motor if you have erratic or slow speed after all drive parts are clean. Do not overlook a defective cassette tape. To be sure, replace it with a new tape.

**Fast forward not working**    Most cassette players will push the idler wheel over to rotate the take-up reel hub. The idler wheel is rotated by friction against a wheel that is connected to the capstan/flywheel shaft. If the tape unit operates normally in Play and slow in Fast Forward mode, suspect slippage within the idler wheel area (Fig. 2-28). Be sure that all drive wheels are clean. With a belt drive fast forward, clean the belt and drive pulley. If the cassette plays slow as well as fast forwarding, clean the motor drive belt and the flywheel area.

On some recorders, the fast forward and rewind are driven by plastic gears. The plastic gears will mesh together when placed into fast forward. The capstan gear drives a larger idler wheel gear that drives another shifting idler gear wheel. The idler gear wheel is moved toward the take-up spindle, which engages two small gear wheels. The plastic gear wheels then rotate when in Fast Forward and Play modes.

These gear-type players will not slip, but might jam if a gear tooth is broken. A misplaced washer or clip might cause gear misalignment and cause the loss of Fast Forward or the Play mode.

**Tape will not rewind properly**    The fast forward and rewind speed is very slow. In the Rewind mode, the idler wheel is shifted when the Rewind button is pressed (Fig. 2-29). Check for worn or slick surfaces on the idler or drive surface area. Clean them with alcohol. Keep in mind that the pinch roller does not rotate in either the Rewind or Fast-Forward mode, only during Record and Play.

**The need to demagnetize tape heads**    Tapes that have been used many times for prolonged periods of time are induced with residual magnetism found in heads, guides, and capstans. A magnetized component (especially heads) anywhere in the tape path will create some hiss and permanent loss of high frequencies on the recorded tape, whether you are recording or just playing the tape. To demagnetize the cassette deck, use a commercially available head demagnetizer. Keep all tapes away from the immediate vicinity of any demagnetizer to avoid accidentally erasing the recorded tapes.

While holding the head demagnetizer away from the tape unit, connect the demagnetizer to an ac outlet and turn it on. Slowly bring the demagnetizer close to each of the surfaces that normally contact the tape. With the demagnetizer still on, slowly withdraw it from the unit (two feet or so), and turn it off.

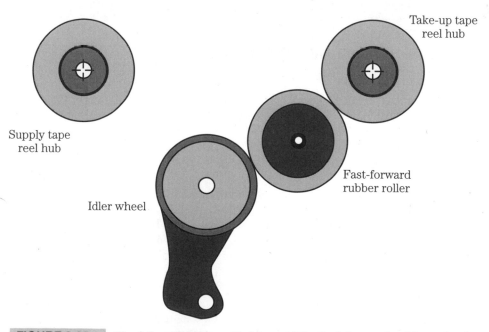

Take-up tape reel hub

Supply tape reel hub

Idler wheel

Fast-forward rubber roller

**FIGURE 2-28**    **The idler wheel is pulled toward the fast-forward rubber wheel, which drives the take-up reel.**

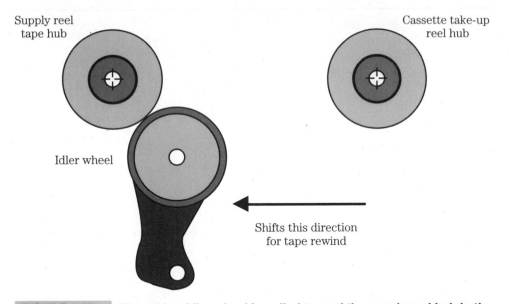

Supply reel
tape hub

Cassette take-up
reel hub

Idler wheel

Shifts this direction
for tape rewind

**FIGURE 2-29**    **The rubber idler wheel is pulled toward the supply reel hub in the tape Rewind mode.**

When servicing the tape deck, do not use any magnetized screwdrivers or other tools near the head or other metal parts that the tape travels around or near. This could magnetize those parts and erase your tape.

## TAPE HEAD CLEANING AND MAINTENANCE

During normal cassette operation, oxide particles are loosened from the tape and build up on the tape head, erase head, capstan shaft, and rubber pinch roller. The erase head is pointed out in Fig. 2-30. The tape player should be cleaned at regular intervals because oxide accumulation can cause distortion and possibly affect tape playback and recording.

Clean the head as follows:

■ Press the Stop/Eject button to open the cassette compartment.
■ Remove the cassette.
■ In some older machines, you might want to press the play lever. The various points that need to be cleaned on a cassette machine is shown in Fig. 2-31.
■ While holding the tape door or lid open, use a long cotton swab to clean heads, capstan shaft, and pinch roller with tape head cleaner or pure isopropyl alcohol (Fig. 2-32).

Both mechanical and electronic deck parts affect sound quality. Today's electronic parts are largely unaffected by dirt and have a very long life. However, the mechanical parts that guide the tape and control its speed for accurate decoding will accumulate dirt and dust. Routine maintenance of these parts will extend the useful life of your recorded tape.

The play/record head (Fig. 2-33) is both mechanical (guides the tape) and electronic (decodes at the head gap). Microscopic dirt caught in the head gap will immediately change the magnetic field and affect the sound quality.

The capstan and pinchroller control the tape speed. As dirt collects, tape slippage and tracking errors occur. The speed becomes erratic and the music sounds slow and warbly.

**FIGURE 2-30**    **The erase head location.**

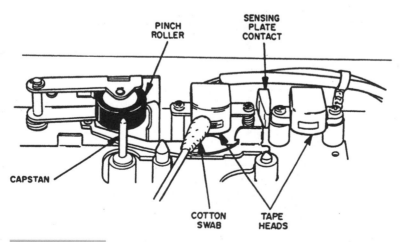

**FIGURE 2-31**    **The various points that need to be cleaned on a cassette machine.**

**FIGURE 2-32**    Use a cotton swab with alcohol to clean the tape record/ play heads.

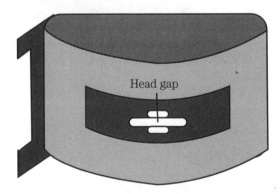

Head gap

**FIGURE 2-33**    The record/play head, showing the head gap detail.

In severe cases, the tape can stick and unwind into the deck mechanical parts. Dirt rarely collects immediately under the moving tape. As the tape rubs across the mechanical parts the dirt shifts above and below the tape's path, or into grooves and gaps, collecting to cause future problems. A primary cause of tape failure is dirt carried on the tape from the deck and wound up under tension in the cassette. Sandwiched between layers of tape, this dirt scratches the metallic oxide, damaging the recorded sound. The "live" sound reproduction is reduced with each play. Regular cleaning prevents dirt from collecting in the deck. Tapes will stay cleaner and last longer.

A habit of routine maintenance and cleaning prevents these problems. Irreplaceable tapes last longer, and you enjoy all of the "live" sound quality that your costly sound system can provide.

**Operation of the Trackmate cleaning cassette**    The Trackmate system has engineered quality cleaning into a single, easy-to-use cassette (Fig. 2-34). Other cleaning cassettes are technically dependent on fabric tape or felt for cleaning. Tapes are ineffective and do not reach where dirt collects, beyond the tape path. Felts touch only a narrow portion of the record/play head, capstan, and pinchroller, missing the erase head, tape guides, and stud posts, leaving them dirty. The Trackmate brushes form fit all these mechanical parts. The 32,000 absorbent, flexible, fibers seek and remove dirt from all of the surfaces and gaps where it collects (Fig. 2-35). Static-control fibers inhibit the attraction of further dust. These high-tech cotton buds have more than 100 times the active cleaning surface area of some earlier products. They automatically clean deck parts from top to bottom, leaving a dirt-free path for the recording tape to safely track around on. Figure 2-36 shows the special cleaning fluid being applied the Trackmate fiber brushes.

## CASSETTE PROBLEMS AND SOLUTIONS OR CORRECTIONS

The following information includes cassette problems that you might encounter and some tips on what to do to solve the problem:

*Symptom:*    Cassette tape will not move.
*Probable cause:*    Clean and/or adjust the control switches.
*Probable cause:*    Motor not running.

**FIGURE 2-34**    **The Trackmate cleaning cassette device.**

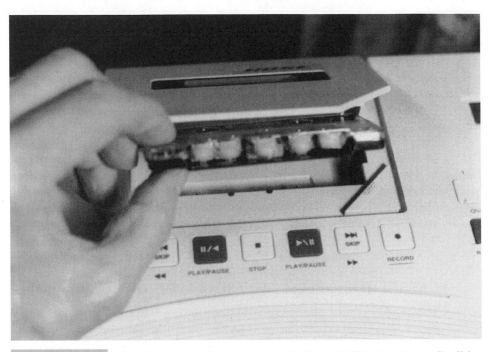

**FIGURE 2-35**    The Trackmate cleaning cassette has 32,000 absorbent, flexible cleaning fibers.

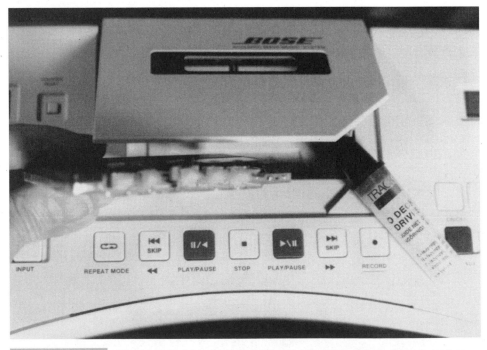

**FIGURE 2-36**    Cleaning fluid is being applied to the Trackmate fiber brushes.

*What to do:* Replace motor.
*Probable cause:* Drive belt worn or broken.
*What to do:* Replace with new belt.

*Symptom:* Tape movement erratic or slow.
*Probable cause:* Motor bearing dry or drive belt worn.
*What to do:* Replace motor or drive belt.
*Probable cause:* Oil or grease on capstan.
*What to do:* Clean capstan with alcohol.
*Probable cause:* Pinch roller dirty or cassette defective.
*What to do:* Clean pinch roller and try a new cassette tape.

*Symptom:* Tape tears or jams.
*Probable cause:* Take-up reel torque is too high.
*What to do:* Adjust or clean turntable clutch assembly.
*Probable cause:* Bent tape guide or misaligned head.
*What to do:* Replace head or readjust.

*Symptom:* Tape will not wind properly.
*Probable cause:* Tape torque is too low.
*What to do:* Adjust clutch subassembly.
*Probable cause:* Clutch arm assembly worn.
*What to do:* Replace clutch arm assembly.
*Probable cause:* Pinch roller out of alignment with capstan.
*What to do:* Adjust pinch roller or replace it.
*Probable cause:* Belt loose or off clutch assembly.
*What to do:* Clean belt and/or replace it.
*Probable cause:* Take-up idler wheel is worn.
*What to do:* Replace idler wheel.

*Symptom:* Tape speed is too slow.
*Probable cause:* Voltage to motor is low.
*What to do:* Check power supply.
*Probable cause:* Drive belt is slipping.
*What to do:* Clean or replace drive belt.
*Probable cause:* Motor stalls.
*What to do:* Replace motor.
*Probable cause:* Pinch roller is dirty.
*What to do:* Clean pinch roller with alcohol.
*Probable cause:* Oil or grease on capstan.
*What to do:* Clean capstan with alcohol.

*Symptom:* Wow and flutter during playback.
*Probable cause:* Cassette pad pressure is too high.
*What to do:* Replace with a new cassette.
*Probable cause:* Pinch roller is dirty or worn.
*What to do:* Clean or replace pinch roller.
*Probable cause:* Oil on capstan or other moving parts.
*What to do:* Clean all of these parts.
*Probable cause:* Capstan shaft is eccentric.

*What to do:*  Replace flywheel.
*Probable cause:*  Tape not following (tracking) in the proper path.
*What to do:*  Check all components and realign the tape path.

*Symptom:*  Fast forward is inoperative.
*Probable cause:*  Fast forward torque is low. Clean or replace fast-forward clutch assembly. Replace spring in fast-forward clutch if pressure is low.
*Probable cause:*  Defective motor.
*What to do:*  Replace motor.

*Symptom:*  Tape will not rewind.
*Probable cause:*  Idler arm damaged.
*What to do:*  Replace idler arm.
*Probable cause:*  Rewind torque is weak.
*What to do:*  Clean fast-forward clutch, idler assembly, and drive reel surfaces from oil, grease, or other impurities. Replace any rubber surfaces that are worn or uneven.
*Probable cause:*  Brake assembly is still in contact with drive reels.
*What to do:*  Adjust, repair, or replace the brake assembly.

*Symptom:*  Rewind speed is slow.
*Probable cause:*  Supply voltage is low or motor is defective.
*What to do:*  Check power supply or install new motor.
*Probable cause:*  Idler is slipping.
*What to do:*  Replace or clean idler wheel.

*Symptom:*  Tape climbs up capstan.
*Probable cause:*  Shaft of pinch roller assembly is bent or loose.
*What to do:*  Replace pinch roller assembly.

*Symptom:*  No audio when playing a tape back.
*Probable cause:*  Defective play/record head.
*What to do:*  Replace or clean play/record head.
*Probable cause:*  Defective power supply or playback amplifier circuits.
*What to do:*  Check the power supply and recorder electronic playback circuits.
*Probable cause:*  Defective cables or cable connections to the power amplifiers or speakers.
*What to do:*  Check all connections and cables. Check power amplifiers for proper operation.

*Symptom:*  No sound, only noise comes from the speakers.
*Probable cause:*  Record/play head open.
*What to do:*  Replace the play/record head.
*Probable cause:*  Open or short circuit in cable to head or faulty plug connection.
*What to do:*  Clean connections and cable or replace cable assembly.
*Probable cause:*  Shielded wire between record/play head and circuitry is pinched, cut, or shorted.
*What to do:*  Replace this shielded wire or repair.

*Symptom:*  Weak playback audio sound.
*Probable cause:*  Dirty play/record head. Check and clean.
*Probable cause:*  Defective amplifier components.
*What to do:*  Check voltages and repair amplifier stages.
*Probable cause:*  Cassette is defective.
*What to do:*  Try a known-good cassette.

*Symptom:* Poor high-frequency audio response on playback.
*Probable cause:* Record/play head is dirty.
*What to do:* Clean the head.
*Probable cause:* Azimuth adjustment is wrong.
*What to do:* Check and correct azimuth adjustment if it is wrong.
*Probable cause:* Record/play head is magnetized.
*What to do:* Demagnetize the head.

*Symptom:* The volume varies on tape playback.
*Probable cause:* Improper pressure of record/play head against the tape.
*What to do:* Adjust for proper head penetration.
*Probable cause:* The tape is not following the proper path.
*What to do:* Check and adjust mechanical components in the tape path.

*Symptom:* Audio is distorted during tape playback.
*Probable cause:* Defective components such, as transistors and ICs, in the playback amplifiers.
*What to do:* Repair audio amplifiers.
*Probable cause:* Defective speakers or connections.
*What to do:* Check for poor speaker lead connections or rubbing voice coils and warped speaker cones. Replace speakers if defective.
*Probable cause:* Record/play head is dirty.
*What to do:* Clean the head and adjust if necessary.

*Symptom:* Tape not being recorded in Record mode.
*Probable cause:* Defective play/record head.
*What to do:* Replace defective head.
*Probable cause:* Defective bias oscillator circuit.
*What to do:* Repair bias oscillator circuit. A typical bias oscillator circuit is shown in Fig. 2-37.

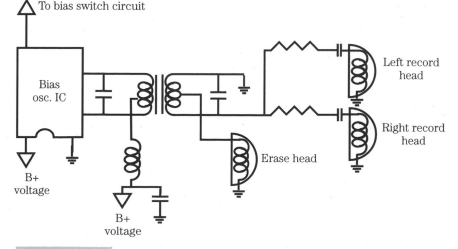

**FIGURE 2-37** A simplified bias oscillator circuit is shown. The bias oscillator signal is fed to the left and right recording heads and to the erase head.

*Probable cause:*  Dirty play/record head.
*What to do:*  Clean the head. Also, the cassette could be defective.

*Symptom:*  No recording be made with microphones.
*Probable cause:*  Defective microphone or microphone plug in jacks.
*What to do:*  Replace microphone or repair the plug-in microphone jacks.

*Symptom:*  Tape cannot be erased.
*Probable cause:*  Defective erase head, dirty erase head, or bias oscillator not working.
*What to do:*  Replace or clean erase head. Repair bias oscillator circuit.

# 3

# AUDIO/VIDEO,

# CD PLAYERS AND

# REMOTE-CONTROL

# OPERATIONS

# How CD and Laserdisc Players Work

To explain CD player operation, this chapter uses a Zenith LDP510 multi laserdisc player. This laserdisc player produces very good quality video and audio.

Figure 3-1 shows the cartridge for a Pioneer audio CD player that uses a six-disc plug-in CD holder. In Fig 3-2, you can see how these CDs swing out to load and also play them. Fully loaded, these cartridges will give you six hours of uninterrupted music in your home or auto.

The Zenith laserdisc and CD player can play back five different kinds of discs:

- *12-inch laserdiscs (LD)*  These can contain up to 120 minutes (60 minutes per side) of high-quality video and digital or analog audio.
- *8-inch laser discs (LD single)*  Plays back up to 40 minutes (20 minutes per side) of high-quality video and digital or analog audio.
- *5-inch compact disc video (CDV)*  This plays up to 5 minutes of high-quality video and up to 20 minutes of digital audio.
- *3-inch compact disc (CD single)*  This plays up to 20 minutes of digital audio.

**FIGURE 3-1**  A six CD cartridge that slides into a Pioneer home or auto CD player.

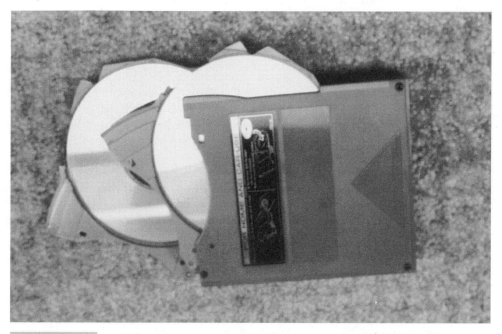

**FIGURE 3-2**    Another view of the Pioneer six CD cartridge, showing how the discs swing out. This cartridge, when full, can provide approximately six hours of music.

This Zenith multidisc player is a remarkably versatile player unit for both video and audio playback modes. Unlike video and audio tape players, these players can quickly find a specific point on the recorded CD.

## SKIP, SEARCH, AND SCAN OPERATION

Now look at how the disc players perform these search and scan operations.

**Chapter/program skip**    Most laserdiscs are divided into segments or chapters, and compact discs are divided into programs. Each program on an audio CD is usually an individual song. With either a laser or compact disc loaded in the CD player machine, pressing the Skip button, forward or reverse, will cause it to skip almost instantly to the beginning of the next or previous chapter or program.

**Chapter/program search mode**    Laserdiscs and CDs with individual chapter or program numbers and descriptions written on the jacket label make it even easier to access specific segments directly. Simply press the Chapter/program key on the remote control and then the desired number of the chapter or program to be played back.

**Disc scan mode**    Both laserdiscs and CDs can be scanned by the Zenith laserdisc player to access a section within a chapter or program. Laserdiscs recorded in CAV (standard play) will show high-speed playback of the video on the TV screen. Compact discs, too, will deliver high-speed audio playback when scanning.

The laserdisc players all use the same operating principles as CDs and CVDs. This Zenith player is capable of playing these discs in addition to laserdiscs.

The recording of the master disc is accomplished (as shown in the description blocks of Fig. 3-3). The video is FM modulated on a 8.5-MHz carrier frequency. Audio signals are

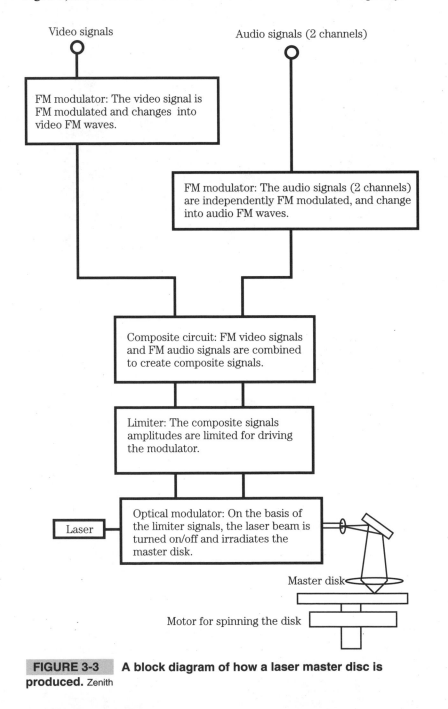

**FIGURE 3-3**    **A block diagram of how a laser master disc is produced.** Zenith

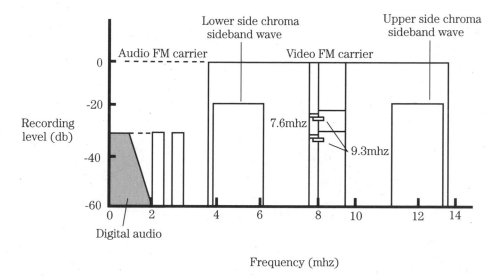

**FIGURE 3-4**    **The frequency spectrum of the laser CD beam.** Zenith

also FM (frequency modulated) on two carrier frequencies at 2.3 MHz and 2.8 MHz. The video and audio signals are combined to create a composite FM signal. This is passed through a limiter before going onto the modulator. In the limiter, the FM signal wave is formed into square waves that are coupled to the modulator to turn the laser beam on and off to create pits in the disc surface.

The master disc is composed of a photoresist, that is exposed by the laser beam, to create pits, in accordance with the video and audio information from the FM carrier wave. The master disc is then used to produce a completed laser vision disc.

The frequency spectrum of a recorded signal is shown in Fig. 3-4. The video frequency spectrum is from 4 to 14 MHz, thus giving it a very wide bandwidth. The chrome signal is comprised of a low frequency and is then frequency converted to produce sidebands as indicated. In this spectrum is a vacant space from 0 to 2 MHz, allowing for digital audio signal recording.

## HOW THE LASERDISC IS MADE

Precision recording and playback of the CD in the very small micron scale is possible because laser beams are used for cutting. The rows of audio and video signals (called *pits*, recorded as bumps on the disc surface) are actually tracks impressed on one side of the CD disc. Each of these tracks contains one video frame.

One picture after another is picked up at the rate of 30 frames per second to produce a full TV picture during playback.

**Disc construction information**   Of the six types of CD discs, the 5-inch size is the most popular today. Refer to Fig. 3-10 for more details on these disc types. The 12-inch disc consists of two one-sided discs (300 mm in size) that are bonded together and are 1.2 mm thick. Program recordings begin at the 110-mm diameter point and end wherever the pro-

(1)Shape and dimension

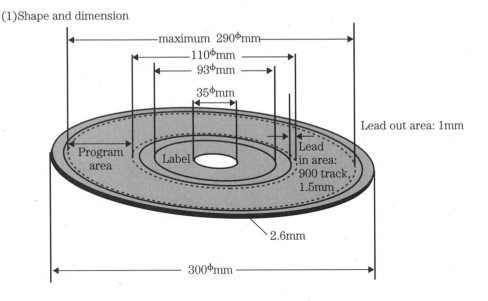

(2)Cross-sectional dimension

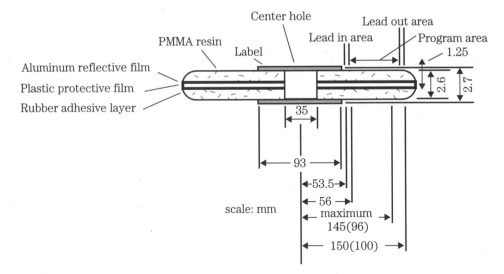

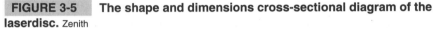

**FIGURE 3-5**    **The shape and dimensions cross-sectional diagram of the laserdisc.** Zenith

gram ends, up to a maximum diameter point of 290 mm. The shape, dimensions and cross-section view of the CD is illustrated in Fig. 3-5. Disc playback systems are classed as either standard (CAV) discs, which play for 30 minutes on one side, or long-playing (CLV) discs, which play for 60 minutes on one side. In either case, playback begins on the inner circumference and ends at the outer circumference.

A lead in about 1.5 mm wide is placed at the inner circumference, which is before the program starting point. The lead in serves as the intro for the program and contains information, such as the trademark of the company that made the disc.

A lead out at least 1 mm wide is located at the outer circumference, which is after the program ending point. It is used to display information, such as the end mark. The installed CD discs rotate clockwise when viewed from the top of the machine.

Standard discs rotate at a constant 1800 RPM (rotations per minute), but because long-playing discs are recorded at a constant linear velocity of 11 m/s, the number of rotations vary with pickup location from the inner circumference (1800 rotations) to the outer circumference (600 rotations). The center hole of the disc is 35 mm in diameter to provide stable support and suppress warping—even when the disc rotates at such high speeds.

**How the disc is constructed**    A resin protective-film coating is applied above an aluminum reflective film deposited on the pit transfer surface of the 1.2-mm thick PMMA base (transparent acrylic resin disc).

Because the protective films are bonded together, the pits of recorded information on finished discs are fully protected by embedding them at least 1.2 mm from the disc surface. Refer to Fig. 3-6 for the construction of the disc's internal structure.

This structure protects against scratches and dust on the signal pickup surface, and prevents scratches and fingerprints from deteriorating the audio or video quality during playback. This is also because the laser beam is precisely focused on the signal (pit) and the disk surface is outside the focal point. Discs have a very long operating life because you can wipe them clean with a soft cloth when they get very dirty. Handling and storage of the CD discs are very easy and convenient.

**Signal (pit) detection scheme**    Now see how the disc signal (pit) and signal pickup is accomplished. The pickup of a laserdisc player extracts signals impressed on the disc for amplification and playback, and generates an output that can be used as audio and video signals for TV picture reproduction.

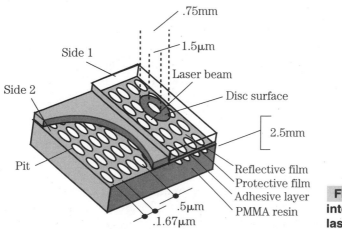

**FIGURE 3-6**    **The internal structure of the laserdisc.** Zenith

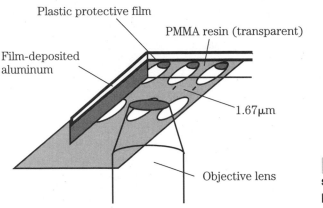

Plastic protective film

PMMA resin (transparent)

Film-deposited aluminum

1.67μm

Objective lens

**FIGURE 3-7    The laser signal detection by the pickup device.** Zenith

Each laserdisc signal is recorded as bump-like pits in varying sizes. The signal recording method is exactly like that for CDs (compact discs), so the method for extracting signals is also the same.

A semiconductor laser emits a pin point of light on a laserdisc. Only light that strikes a pit and is reflected is picked up and converted to an electrical signal. This operation is depicted in Fig. 3-7. The slight distance maintained between the pickup and the signal surface prevents damage to the pickup and disc. These are the advantages of the noncontact-type laserdiscs.

**Optical pickup and detection via the pit signal**    Almost all light emitted from the semiconductor laser is reflected at locations without the pits. A reflected light differential (power) is generated at the pit locations because only a portion of the light is reflected.

Because the length of each pit differs according to the impressed information, the reflected light differential (power) based on the varying length is converted to an electrical signal by the photodiode. Eventually, it becomes the audio and video signals, as shown in Fig. 3-8.

Light shone on the disc and reflected back just as it would return to its place of origin during the process of signal extraction, so an extraction method that can identify the light is required. For this job, a half mirror is used because it reflects 50% and passes the other 50%.

Besides the route just described, the light also passes through the grating, collimator lens, and objective lens. Each of these items are designed to control the direction in which the light advances and then to assign the correct signal to the pit.

**The laserdisc pits**    The size of a disc pit that represents a recorded signal is extremely small. An enlarged view of these pits are shown in Fig. 3-9.

A standard CD contains approximately 12 to 15 billion pits on one side. To get some perspective on this number, you can think of it as roughly equivalent to the number of brain cells in an adult person. The large 12-inch disc would contain much more information.

These tiny pits lined up on a single circuit of the disc is called a *track*. One track contains information for a single picture or screen full on a TV. Two fields, like that on a TV, are formed from 30-frame screens every second.

Movies are 24 frames and one laserdisc track equals one frame. Consequently, one side of a laserdisc records 54,000 tracks. Because 30 tracks form a one-second image, one side of a disc records 30 minutes of video.

**Various types of CDs**  The following types of discs can be played back on laser players nearly all discs are LD, CD, or CDV compatible. Let's now look at the specifications of these various laserdiscs.

***Standard (CAV) discs***  These standard discs have a constant angular velocity. A disc spins with a constant rotational speed of 1800 RPM. Playback time on one side of a 12 inch disc is 30 minutes, recording a maximum of 54,000 frames of picture information. As

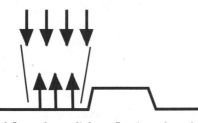

(a) Laser beam light reflection when there is no pit on the signal surface (almost all the light is reflected)

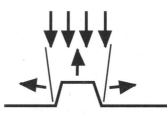

(B) Laser beam light reflected by a pit

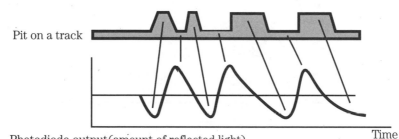

Pit on a track

Photodiode output(amount of reflected light)

Time

**FIGURE 3-8**    **Signal extraction by the optical pickup device.** Zenith

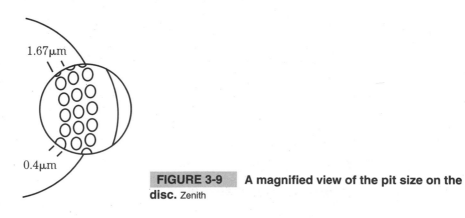

1.67μm

0.4μm

**FIGURE 3-9**    **A magnified view of the pit size on the disc.** Zenith

the disc spins once, it turns each picture frame, which are each provided with frame numbers from 1 to 54,000.

***Long play (CLV)*** These discs have a constant linear velocity. The rotational speed changes accordingly, when the signal in the inner circumference is being read, it spins at 1800 rpm. When the signal in the outer circumference is being read, it spins at 600 rpm. The playback time for one side of a 12-inch disc is 60 minutes. The playback elapsed time is recorded on the disc from the beginning.

***Compact disc with video (CDV)*** Pictures associated with digital sound are recorded on the outer tracks (video part) five minutes long. Digital audio on the inner tracks (audio) 20 minutes: A normal playback begins at the video part through to the audio part.

The main component of this CD equipment is the mechanical operating portion. This part of the CD player is broadly divided into the pickup carriage portion and the mechanical subchassis section that handles the loading and motor elevation operations. This CD player section is shown in Fig. 3-11.

### How the pickup carriage functions

■ The pickup carriage, which features all tilt mechanisms, including a tilt sensor and tilt motor, is driven in the feed direction along the feed rack. A drawing of this pickup carriage unit is shown in Fig. 3-12.

■ This pickup carriage is equipped with a mechanism to adjust the tilt sensor mounting angle (screw) so that the sensor is constantly vertical in the radial direction with respect

Types of discs

The following types of discs can be played back on laser players because nearly all discs are LD, CD, or CDV compatible

| Disc | CD | | CDV | LD | | |
|---|---|---|---|---|---|---|
| Disc size | CD single<br><br>3-inch | 5-inch | 5-inch<br><br>B<br>A | LD single<br><br>8-inch | 8-inch | 12-inch |
| Maximum recording time | 20 minutes (one side only) | 74 minutes (one side only) | 25 minutes (one side only) | CAV: 14 mins.<br>CLV: 20 mins.<br>(one side only) | CAV: 28 mins.<br>CLV: 40 mins. | CAV: 60 mins.<br>CLV: 120 mins. |
| Recorded contents | Audio only: digital audio | | A: Video&Digital audio (5 min.)<br>B: Digtal audio (20 min.) | Video & digital / analog audio | | |
| Indication symbol | CD SINGLE | disc COMPACT DIGITAL AUDIO | CD VIDEO | CD VIDEO | LASER DISC | |

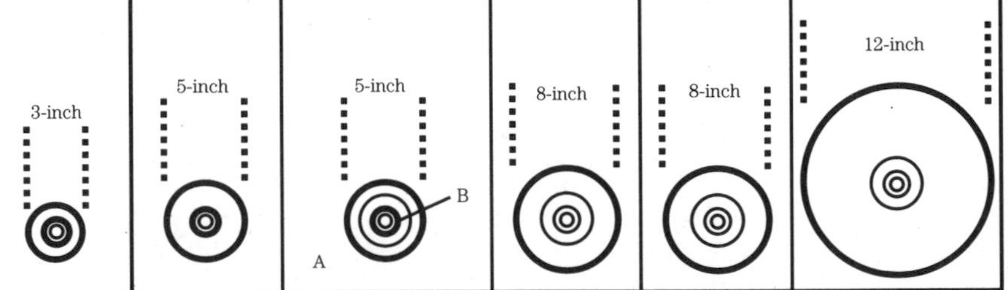

**FIGURE 3-10**    **The various types and sizes of laserdiscs.** Zenith

## MECHANICAL PRINCIPLE

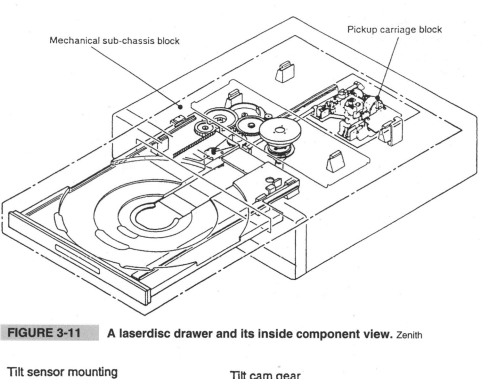

**FIGURE 3-11**    **A laserdisc drawer and its inside component view.** Zenith

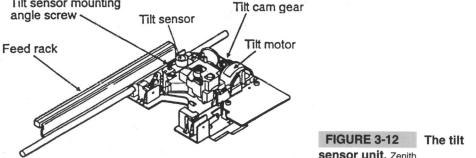

**FIGURE 3-12**    **The tilt sensor unit.** Zenith

to the disc position. The tilt cam gear, which controls tilt, is equipped with an overtilt mechanism to lower the position of the pickup when tray loading, and a limiter mechanism to prevent the tilt cam gear from over rotating.

■ The pickup base to which the pickup is attached is adjusted in the normal direction by a screw to maintain positional accuracy of the pickup in the normal adjustment direction.

### How the mechanical subchassis works

■ The loading mechanism, spindle motor elevation mechanism that clamps the disc, and the feed mechanism for the pickup carriage are located on the same chassis. All of these operations are handled by a single loading motor.

- The loading mechanism uses an auto loading system so that a light push on the tray or a press of the Open/close button triggers the loading motor to pull in the drawer. On some PCs, the CD drawer is pulled in (loaded) as the program is being run.
- The clamp mechanism raises the spindle motor to clamp a disc between the turntable and the clamper mounted on the subchassis.

Protection, setting mode, playback, and disc ejection of each disc is performed by each mechanical sequence on the subchassis that has just been described.

**Mechanical tray operations**   Refer to Fig. 3-13. The tray has a guide groove on the right side to engage the resin guide on the subchassis, as well as a drive transmission rack for horizontal movement that engages the slide on the right bottom surfaces and drive gears on the left bottom surfaces.

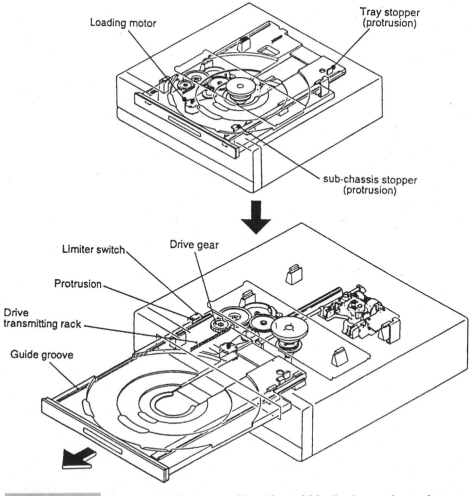

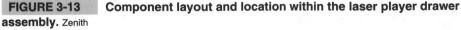

**FIGURE 3-13**    **Component layout and location within the laser player drawer assembly.** Zenith

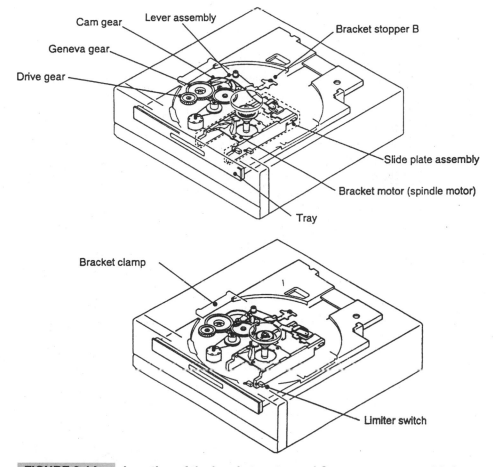

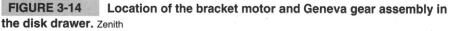

**FIGURE 3-14**    **Location of the bracket motor and Geneva gear assembly in the disk drawer.** Zenith

A protrusion at the rear of the tray serves as a stopper. When the tray is finished unloading, the stopper contacts the protrusion on the subchassis to prevent the tray from moving too far forward.

The end of slide tray unloading operations is detected when the protrusion at the left rear of the tray presses the limiter switch mounted on the subchassis side. In other words, the loading motor is off when the tray is unloading, but a slight press on the tray breaks contact between the protrusion on the tray and limiter switch. This causes the loading motor to turn on, and the tray is automatically pulled into the player. Normal tray operation can be performed by the player operating buttons.

**The disc loading operations**    Refer to Fig. 3-14 for how the disc loading is accomplished. The loading mechanism is made up of the following items:

- Tray.
- Slide plate assembly.

■ Lever assembly.
■ Bracket stopper "B."
■ Bracket motor (spindle motor).

When the tray is housed in the player, the Geneva gear that engages the cam and drive gears locks to prevent the tray from moving back and forth. Grooves at the rear of the cam gear engage the lever assembly pin, causing the lever assembly to rotate. The other end of the lever pin pulls the slide plate assembly to slide the assembly horizontally.

Three angled grooves on the inner side of each side plate of the slide plate pair on the assembly engage protrusions from either side of the bracket motor (spindle motor). Two shafts set up on the rear surface of the subchassis restrict the bracket motor to the vertical direction and align it in the horizontal direction to ensure that the bracket motor operates vertically along the angled grooves. When the bracket motor is firmly pressed against the rear surface of the subchassis, the slide plates cut off the limiter switch, and the movement stops. This engages the lever assembly and bracket stopper B, which locks the lever assembly into position.

The spindle motor rises, causing the turntable to raise the clamper and the disc in the tray. The clamper clamps the disc with magnetic force between the bracket clamps and plate spring force.

**Pickup carriage operation**   The pickup carriage operates as follows in each mode of the loading operation. Refer to Fig. 3-15 for the carriage operation.

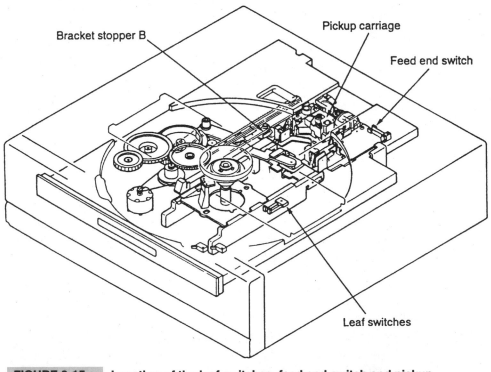

**FIGURE 3-15**   **Location of the leaf switches, feed end switch and pickup carriage.** Zenith

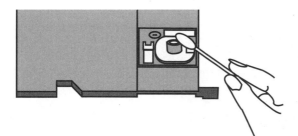

**FIGURE 3-16** How to use a cotton swab to lightly wipe the lens opposite the pick-up carriage. Zenith

**Tray out/tray in** During this time, the pickup carriage is positioned farther out than the outer periphery of the disc. The feed end switch is pressed and the carriage is locked in the feed direction by bracket stopper B. The pickup reaches overtilt by moving to the maximum tilt operating range, lowering the pickup so that it does not interfere with the tray.

**Spindle down/spindle up** During this time, tilt mechanism operations positions the pickup nearly horizontally. Overtilt and horizontal tilt are determined by detecting the rotation position of the cam gear. This position is detected by the reflector plate mounted on the rear side of the cam gears that operate tilt, and by two photo reflectors, which are mounted on the subchassis directly across from the reflector plate.

**Spindle-up complete-feed operations** When the spindle is completely up, the pickup carriage is released by the movement of the bracket stopper B. The feed rack drive gears, which were locked, rotate and the pickup carriage moves toward the inner periphery. The feed rack then engages the gears closest to the loading motor for feed operations with minimum backlash in the range applicable for actual disc playback.

The inner periphery of each disc is detected when the pickup carriage turns the tandem leaf switches mounted on the subchassis on and off. This way, the positions necessary for playback of each disc, such as the lead in for CDs and DCVs, the video part of CDVs, and the lead in for LDs, is detected. During playback, the tilt sensor signal keeps the tilt mechanism operating as an ordinary tilt servo, and keeps the pickup horizontal to the disc surface. It also ensures stable playback—even with extreme warping by setting the tilt fulcrum point closer to the inner side of the pickup.

**Pickup lens cleaning of the laserdisc player** Depending on the environment where the equipment is used, the pickup lens will become dirty (dirt/dust) after an extended period of time. Dirt will deteriorate picture and sound quality during playback, as well as destabilize the playback. When this happens, clean the lens:

1 Remove the laser player cabinet.
2 Use a new cotton swab to lightly wipe the lens located opposite the pickup carriage section. Wipe two or three times in a spiral moving from the center to the outer periphery (Fig. 3-16).

■ If the pickup carriage is not at the feed end (far peripheral standby) position, turn on the power, position the pickup carriage at the feed end, and turn off the power before cleaning.
■ Do not leave the laser player out of the cabinet for any extended period of time.

■ If you accidentally get dirt on the lens, such as fingerprints during cleaning, wipe with a cotton swab. If the dirt does not come off by wiping, place a small amount of isopropyl alcohol on a new cotton swab and wipe two or three times in a spiral motion, moving from the center to the outer periphery (edges).

■ Do not use any type of alcohol besides isopropyl alcohol. Other types of alcohol might damage the lens.

■ Do not wipe the lens forcefully because this might cause scratches on the lens.

# DVD Discs

These newer dual-layer high-density CDs, called *DVDs*, produce good-looking video and great sound quality. However, the entertainment industry does not know exactly what DVD stands for. It could be "Digital Video Disk" or "Digital Versatile Disk," but no one seems to know for sure at this writing.

Some of the first users of these DVD disks will be the group of video connoisseurs that have for many years been buying the 12-inch laser discs that can match the DVDs video and audio quality. These DVD disks hold much more information, cost less, and are easier to use and store than the 12-inch laser discs.

## DVD TECHNOLOGY

With compression technology you can squeeze as much as 17 gigabytes of data onto a disk the size of a 5-inch CD, which can only hold about 650 megabytes of data. The pits that actually make up the present standard CD, means they can be jammed closer together and are read by a more-accurate laser beam. And a more important feature is that the material that is recorded is compressed and requires a lot less space.

On the CDs now being used, you have to turn a disc over to play the other side or have one laser on top and another laser on the bottom of the discs. On these new laser discs, so as not to turn them over for a long movie, two separate program layers are recorded on one side of the DVD. When playing this type disc, the laser will first focus on one layer and then onto the other layer without any interruption. Some discs use double layers on both sides of the disc for an even longer playing time. This dual-layer technique is illustrated in Fig. 3-17.

## LASER LIGHT AND LASER DIODE INFORMATION

*Laser* is an acronym for *light amplification by simulated emission of radiation*. The laser device produces coherent radiation in the visible light range. The radiation of the laser beam is very narrow and does not spread out over a great distance, unlike light from a standard light bulb. Thus, the beam can be controlled to pin-point accuracy and can also be modulated. There are many different types and power ranges of laser devices.

The laser diode is a type of semiconductor device that emits a coherent light beam. The diode has an internal reflection and reinforcement capabilities. The laser diode is the size of a crystal of table salt.

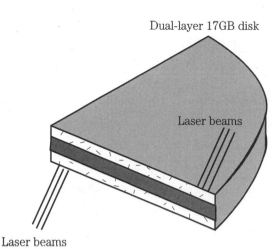

Dual-layer 17GB disk

Laser beams

Laser beams

**FIGURE 3-17**    **A cut-away view of a dual-layer disc. Compression squeezes as much as 17 gigabytes of data into a disc the size of a 5-inch CD.**

Laser diodes are also used as emitters in short- and long-range fiberoptics cable systems and sensors in your CD players. The typical laser diode uses 100 to 200 mW to power it and develops about 3 to 5 mW of output power to produce one very thin light beam.

## CD PLAYER SECTIONS

Now review and summarize the various sections of a typical CD player.

**The power supply**  The CD player does not require a large-current power supply and the portable units use batteries. Most late-model CD players use the switching type of power supply and it is a sealed unit and is not recommended for repairs. Most develop three levels of voltage from the power supply. For the logic power, the $V_{cc}$ is +5 Vdc. Analog and motor disc drive players use ±15 V).

**The optical deck section**  All of the components located on the optical deck are used to load and spin the CD and some type of motor-driven loading device. On portable units, the drawer usually has to be pulled or pushed manually.

***Some probable troubles***  Troubles in the optical deck could be an loose or oil slick belt that causes the drawer not to open or close. It might also stick and not go through its cycle. Also, look for worn or damaged gears and dirty control switches.

## THE ELECTRONICS PC BOARD

Problems in this board call for a professional electronics technician to find and repair any problem. This PC board also has many adjustment controls that should not be changed unless you are a qualified CD service technician.

## THE DISC MOTOR

The disc motor is called the *spindle motor* and it spins the CD disc.

**Some probable troubles**   A defective motor with an open or shorted motor winding. Worn, dirty, or dry motor bearings. If your unit has a drive belt, it could be dirty, worn, or loose.

## THE SPINDLE PLATFORM TABLE

The CD disc sets on this spindle turntable.

**Some probable troubles**   A dry or worn spindle table bearing can cause the CD to wobble and cause intermittent loss of the playback video. The spindle could also be bent and the table could be dirty. Lubricate the bearing and carefully clean the turntable and all other components in this area.

## THE SLED MECHANISM

The sled is the device where the optical pickup is mounted. The optical device is mounted on the sled so that it can be moved across the disc to play and/or locate specific portions of data. The sled is on guide rails and is moved by a worm gear and a linear motor. It works very much like the hard dive in a PC.

**Some probable troubles**   Check for dirt and grease on the slide rails. They might be gummed up, have damaged gears, or need lubrication.

## THE PICKUP MOTOR

The optical pickup is mounted on the sled which zips across the disc to access the various sections for playback. It uses a dc motor that can be belt or gear driven.

**Some probable troubles**   The sled motor could be defective and not run. The gears might be worn or jammed or dry. The belt might be broken or very loose.

## THE DISC CLAMPER

In most machines, the clamper is a magnet on the opposite of the disc. The clamper magnet keeps the disc from slipping on the spindle platform.

**Some probable troubles**   The clamper will not completely engage; when this occurs, the disc will slip on the spindle. This problem is usually caused by a mechanical trouble in the drawer-closing components.

## THE OPTICAL PICKUP UNIT

You can compare this device to a record player, which uses a needle (stylus) to contact the grooves in the record to play audio. In this case, the optical pickup reads the information encoded on the CD. This unit includes the laser diode, optical device, focus and tracking components, and the photodiode assembly. All of the optical pickup components are mounted on the sled, which is connected to a servo and reading processing electronics with flat, flexible cables.

**Most common trouble** Because this flexible cable is moving at all times during laser player operation, this component can have a high failure rate. This flat ribbon will develop very small cracks and can cause some very unusual intermittent problems. Replace the cable if it has these problems.

# Some Common CD Player Problem Areas

- *The CD player will operate OK, then stop in same location of the disc* Your unit might have a transport lock screw. Check to be sure that it is in the Operate position.
- *Drawer loading problem because of belts* Belts loose, very worn and slipping, oil on them and cracked due to age.
- *Poor video or audio* The optics are dirty. Clean the lens, prism and turning mirrors.
- *Mechanical section not working properly* Parts dirty and need to be cleaned and lubricated, grit or sand in the moving parts.
- *Broken parts* This includes brackets, mountings and gears.
- *Intermittent interlock or limit switches* They are dirty and need to be cleaned or need adjustment
- *Intermittent operation because of poor electrical connections* Check poor cracks or poor solder joints on the PC board, poorly contacting connectors, or broken flex cable trace leads.
- *Motors* The winding could become open or shorted. The motor bearing can become worn, loose, or dry and need lubrication.
- *Electronic servo problem* The servo requires focus, tracking, or PLL (phase-lock loop) adjustments.
- *Laser defect* The laser diode might be dead, weak, or not be receiving correct dc power.

The laser diode has a very low failure rate.

- *Photodiode array problem* The diode might be weak, defective, or have shorted segments. Also look for faulty heat-sensitive components.

# Checking and Cleaning the Laser Player

If the CD laser player is operating erratically, you need to check the drawer components and sled drive unit to see if it needs to be cleaned and lubricated. Also check and clean the objective lens.

Carefully clean the lens because it is very delicate. Start by blowing out any dust or dirt around the lens. Then clean the lens with special lens cleaners. It is made of plastic, so do not use any strong solvents. Pure isopropyl alcohol is also effective for cleaning the lens.

A CD lens-cleaning disc is not as effective because it does not remove grime and grease, and can sometimes cause the performance to be worse.

Check the spindle bearing because this can cause noise in the audio. There should not be any play in the CD platform.

Check the drawer mechanism for smooth operation. Clean and lubricate if it needs it. Check the belts to see if they are worn or loose. Also check the motor and gears for proper lubrication.

Check out the various components in the sled drive that moves the pickup device. Look for worn belts, worm gears, and slide bearings. They might need to be cleaned and/or lubricated.

# More CD Player Checks

If the CD player is completely dead, make the following checks. For a dead player, check out the power supply, power cord, and ac plug first. Locate the power-supply section and check for blown fuses. The power supply usually is +15 and −15 Vdc. Use a voltmeter to see if the power circuits are producing this voltage. Some CD players have an interlock switch. If this switch goes bad, it can cause the player to operate intermittently. It might only need to be cleaned or adjusted.

Check for any loose or poor connections. These plugs and connections are usually very small. Cleaning, pushing, and resetting them can clear up many strange problems in these players. Also check out any of the ribbon cables by moving and flexing them and observing any intermittent problems that might occur.

The switch contacts might become dirty because of oil, grease, or oxidation. This will cause the limit or interlock switches from making good contacts. When this occurs, the CD player might not accept a disc, stop at random intervals, or close the drawer in an intermittent manner. Use a contact cleaner and burnish the switch contacts.

# How Remote-Control Systems Work

This section covers various remote-control systems, troubles that might develop, and what to do when they don't work.

The first TV remote "wireless" controls used ultrasonic frequencies in the 35- to 45-kHz range. Some of the Zenith remotes used hammers (clickers) to strike tuned metal rods (usually four) in the hand unit to produce (ring) one ultrasonic control frequency. With this set up, you could rattle a key chain and make the TV go off/on or change channels. These early model remotes produced only four to eight analog frequencies. Some later-model Magnavox remotes generated ultrasonic control pulses and would have 10 or more remote control functions.

Some modern-day remote controls are shown in Figs. 3-18 and 3-19. Remote units now control TVs, VCRs, camcorders, stereo audio units, cable TV converter boxes, DBS satellite receivers, laser CDs, and much more. Of course, many multi-type remote controls will operate several different devices at the push of a button.

**FIGURE 3-18**    Some current model remote-control hand units for TVs and VCRs.

**FIGURE 3-19**    More current remote units used for DBS satellite receivers and cable TV control boxes.

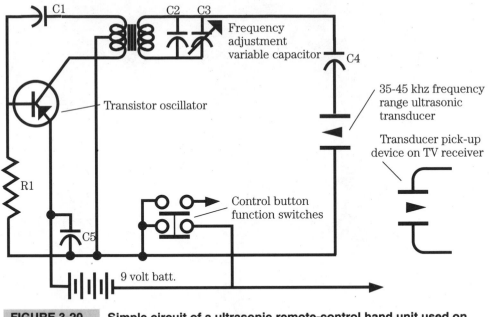

**FIGURE 3-20**    **Simple circuit of a ultrasonic remote-control hand unit used on older-model TVs.**

## THE ULTRASONIC REMOTE TRANSMITTER

The ultrasonic remote control was used in the older TVs and very few of these systems are now in use. These remotes transmitted on an ultrasonic frequency range of 35 kHz to 45 kHz. A few models would generate 10 to 15 control pulses that could be decoded in the TV receiver for more logic control functions. A simplified circuit drawing of an ultrasonic remote-control unit is shown in Fig. 3-20.

## THE INFRARED (IR) REMOTE-CONTROL TRANSMITTER

Modern remote controls use an infrared (IR) carrier frequency that is pulse-code modulated. The carrier frequency is approximately 35 kHz to 55 kHz. The pulses sent out are multiple cycles of usually 20 bits each that modulate the carrier. The logic coding is different for various devices, so only a particular remote will operate a device. However, a universal remote can be reprogrammed for many different kinds of devices and multipurpose remote control units are also supplied with many TVs and VCRs. A block diagram of an infrared digital remote-control transmitter is shown in Fig. 3-21.

Figure 3-22 shows the infrared remote receiver located within a TV, VCR, etc. The IR signal is picked up by an IR diode sensor on the front panel of a TV and is amplified and pulse decoded. The pulse codes are then sent to a remote-control microprocessor IC, which then sends control voltages to various parts of the TV circuits to control the set's operation, such as power on/off, volume control, etc. Figure 3-23 shows a typical color TV remote transmitter.

**What to do when the remote control will not work**  Remote-control units do not fail very often, unless they have been dropped, thrown around, or dunked in some kind of liquid. If your remote equipment does not work, make these quick checks:

**1** If your TV (or other device) has a master on/off/manual/remote switch, be sure that it is in the Remote position.
**2** If it has a multi-function control, be sure that it is not in the VCR function mode when you are trying to operate the TV.

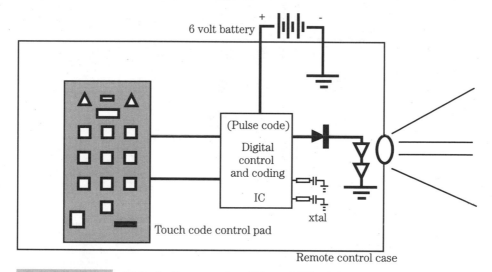

**FIGURE 3-21**  **A block diagram of an infrared (IR) digital remote-control transmitter.**

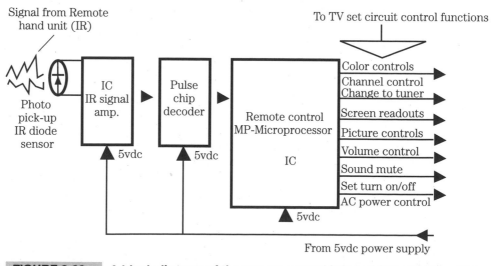

**FIGURE 3-22**  **A block diagram of the remote-control infrared (IR) circuits within the TV.**

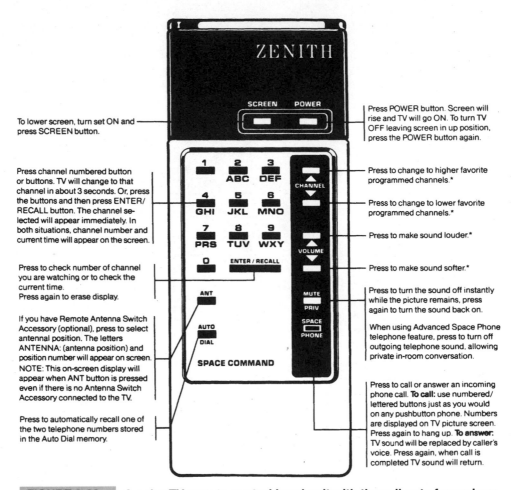

To lower screen, turn set ON and press SCREEN button.

Press POWER button. Screen will rise and TV will go ON. To turn TV OFF leaving screen in up position, press the POWER button again.

Press channel numbered button or buttons. TV will change to that channel in about 3 seconds. Or, press the buttons and then press ENTER/RECALL button. The channel selected will appear immediately. In both situations, channel number and current time will appear on the screen.

Press to change to higher favorite programmed channels.*

Press to change to lower favorite programmed channels.*

Press to make sound louder.*

Press to check number of channel you are watching or to check the current time.
Press again to erase display.

Press to make sound softer.*

If you have Remote Antenna Switch Accessory (optional), press to select antennal position. The letters ANTENNA: (antenna position) and position number will appear on screen. NOTE: This on-screen display will appear when ANT button is pressed even if there is no Antenna Switch Accessory connected to the TV.

Press to turn the sound off instantly while the picture remains, press again to turn the sound back on.

When using Advanced Space Phone telephone feature, press to turn off outgoing telephone sound, allowing private in-room conversation.

Press to automatically recall one of the two telephone numbers stored in the Auto Dial memory.

Press to call or answer an incoming phone call. **To call:** use numbered/lettered buttons just as you would on any pushbutton phone. Numbers are displayed on TV picture screen. Press again to hang up. **To answer:** TV sound will be replaced by caller's voice. Press again, when call is completed TV sound will return.

**FIGURE 3-23**   **A color TV remote control hand unit with the call-outs for various functions.** Zenith

3  If you have a universal remote unit, it might have become deprogrammed or programmed for the wrong model. Also, if the battery has been replaced, it will need to be reinitialized.

4  Speaking of batteries, you should now be sure that the battery is in good condition, with clean, corrosion-free contacts.

   Install a new known-good battery or use a dc voltmeter for a voltage check. The battery should be checked while under load in the remote unit. With the voltmeter connected to battery terminals, press any button and see if the voltage drops more than 10%. If it does replace the battery with a new one. You can also use a battery tester meter because it puts the correct load on the battery for a valid test.

5  Be sure that you are using the correct remote unit because they look alike and some homes might have six or more various remote controls.

6  The problem could be in the TV or VCR that you are trying to operate. See if the TV or VCR will operate manually. If it does, try another remote, such as a universal one to determine if it is the remote control or TV/VCR.

**7** To find out if your remote unit is transmitting an IR signal, you can use an IC detector card (Fig. 3-24). The card will show a red pulsing spot on the card if the control is transmitting. However, this does indicate if it is sending out the correct pulse codes. These cards are available at electronic parts stores.

**8** If the control unit has gotten wet, you might still be able to save it. As soon as possible, flush the unit in clean water and if you can take the case apart, use a hair dryer to completely dry out the case and circuit board. Then clean the battery contacts and install a new battery. It might keep on clicking. It's worth a try and you might not have to buy a new remote control.

## UNIVERSAL REMOTE-CONTROL DEVICE

The Radio Shack universal remote control (Fig. 3-25) can replace many types of standard remote controls. These units are preprogrammed and do not have to "learn" commands from the original remote. Just "tell" this unit what remote control you want to replace (by

**FIGURE 3-24**    Testing a remote IR transmitter with an IR detector card.

**FIGURE 3-25**    **The Radio Shack 3-in-1 universal remote control. This can be programmed for many brands of TVs, VCRs, DBS receivers, cable boxes, etc.**

entering the three-digit codes shown for many brands (Fig. 3-26) and the 3-in-1 universal remote control does the rest.

**How to program the universal remote**    Follow these steps to set up the 3-in-1 remote:

**1** Check or install new batteries before programming the unit.

> Do not place objects on top of the remote control after you install the batteries. Something could press down the keys and reduce the battery life.

**2** Refer to the device codes in Fig. 3-26 and write down this information.
**3** Press the device key for the remote that you are replacing (TV, VCR, or CBL).
**4** Press down and hold PROG until the red indicator blinks, and continue holding it down as you enter the 3-digit set brand code.
   For example, to replace a Panasonic TV's remote control (code 051), you would press:

```
TV-PROG 0 5 1
```

**5** When the LED indicator blinks twice, release PROG.
**6** Point the remote 3-in-1 at your device you want to control and press Power. The device should turn on or off, if it was already on.

Repeat steps 2 through 6 for any additional devices.

The punch-through feature is automatically turned on for the TV's volume and mute controls. This means that when you select CBL and press one of the volume buttons or the Mute button, the remote actually sends the codes to the television and not to the cable converter box. If you want to use your cable converter's volume and mute controls, disable the punch-through feature for these buttons.

If the remote does not operate your device, try other codes listed in Fig. 3-26 for your brand of TV, VCR, or cable box.

The red indicator at the top of the 3-in-1 remote flashes twice when you enter a code that it recognizes. (This does not mean that it is the right code for your device, however.)

When the universal remote's range decreases or the remote operates erratically, replace with new batteries.

---

**Latest Code Lists for Programming the Remote Control**

This sheet contains the latest brand and code information.
See your User's Manual for instructions on programming the remote to control other brands of equipment.

### TV Codes

| Brand | Code |
|---|---|
| Akai | 002 |
| Anam National | 038 |
| AOC | 011,019,027,088 |
| Candle | 011,027,033 |
| Citizen | 011,027,033,064 |
| Colortyme | 011,027,084 |
| Concerto | 011,027 |
| Contec/Cony | 036,057,040,042,064 |
| Craig | 064 |
| Curtis Mathes | 000,011,015,027,037 |
| CXC | 064 |
| Daewoo | 011,019,027 |
| Dayston | 011,027 |
| Electrohome | 006,011,014, 027,058,061,068 |
| Emerson | 011,026,027,028, 029,030,031,032,037,042,053, 064,065,067,075,076,078,079 |
| Envision | 011,027 |
| Fisher | 017,021,039,041 |
| Funai | 064 |
| GE | 000,008,009,011,012, 027,038,068,086,089,091 |
| Goldstar | 003,004,006, 011,019,027,037,050 |
| Hallmark | 011,027 |
| Hitachi | 009,011,027,036, 057,040,047,063,080 |
| Infinity | 013 |
| JBL | 013 |
| Jensen | 011,027 |
| JVC | 012,024,056, 057,040,048,051,074 |
| Kawasho | 002,011,027 |
| Kenwood | 006,011,014,027 |
| Kloss Novabeam | 035,043 |
| KTV | 078 |
| Loewe | 013 |
| Luxman | 011,027 |
| LXI | 013,018,021,023,054 |
| Magnavox | 006,007, 010,011,013,016,027,033, 035,043,049,066,087,089 |
| Marantz | 013 |
| Marantz | 011,013,027,069 |
| MGA | 006,011,014,019, 022,027,041,056,061,068 |
| Mitsubishi | 006,011,014,019, 022,027,041,055,056,061,068 |
| MTC | 011,019,027 |
| Multivision | 081 |
| NAD | 018,023 |
| NEC | 011,014,019,027,038,084 |
| Panasonic | 012,013,058,086 |

| Brand | Code |
|---|---|
| Penney | 000,008,011,019, 027,040,068,077,086,088 |
| Philco | 006,007,010,011,013, 016,019,027,033,035, 057,058,043,087,089 |
| Philips | 002,006,007,010, 011,013,016,033, 035,057,038,043,066,073 |
| Pioneer | 011,027,045,062,093 |
| Portland | 011,019,027,037 |
| ProScan | 000 |
| Proton | 011,027,057,072 |
| Quasar | 012,038,092 |
| Radio Shack | 000,021,025, 056,057,059,064,078 |
| RCA | 000,006,011,019, 027,034,038,044,046,088 |
| Realistic | 021 |
| Sampo | 011,027 |
| Samsung | 006,011,014, 015,019,027,036,057,077 |
| Sanyo | 017,021,039, 056,057,058 |
| Scott | 028,037,064 |
| Sears | 000,006,011,014, 017,018,021,023,027, 039,040,041,051,071,083 |
| Sharp | 011,020,025,027, 037,052,053,059,060 |
| Sony | 002 |
| Soundesign | 011,027,033 |
| Sylvania | 006,007,010, 011,013,016,027,033, 035,043,049,066,087,089 |
| Symphonic | 064,076 |
| Tatung | 038 |
| Technics | 012 |
| Techwood | 011,027 |
| Teknika | 011,019,027, 033,036,057,040,066 |
| Telecaption | 090 |
| TMK | 011,027 |
| Toshiba | 018,021,023, 040,071,077,085 |
| Universal | 008,009 |
| Victor | 051 |
| Victech | 019,027 |
| Wards | 000,005,006, 007,008,009,010,011, 013,019,025,027,028, 035,043,059,066,076,082,089 |
| Yamaha | 006,014,019,027 |
| Zenith | 001 |

### Cable Box Codes

| Brand | Code |
|---|---|
| ABC | 022,046,053,054 |
| Anvision | 007,008 |
| Cablestar | 007,008 |
| Diamond | 056 |
| Eagle | 007,008 |
| Eastern Int. | 002 |
| General Instuments | 046 |
| GI 400 | 004,005,015, 023,024,025,030,036 |
| Hamlin | 003,012,013,034,048 |
| Hitachi | 037,043,046 |
| Jerrold | 004,005,015,023, 024,025,030,036,045,046,047 |
| Macom | 057,045 |
| Magnavox | 007,008,019,021, 026,028,029,032,033,040,041 |
| NSC | 009 |
| Oak | 001,016,038 |
| Oak Sigma | 016 |
| Panasonic | 003,027,039 |
| Philips | 007,008,019,021, 026,028,029,032,033,040,041 |
| Pioneer | 018,020,044 |
| RCA | 000,027 |
| Randtek | 007,008 |
| Regal | 003,012,013 |
| Regency | 002,033 |
| Samsung | 044 |
| Sci. Atlanta | 003,022,035 |
| Signature | 046 |
| Sprucer | 027 |
| Starcom | 046 |
| Stargate 2000 | 058 |
| Sylvania | 011,059 |
| Teknika | 006 |
| Texscan | 010,011,059 |
| Tocom | 017,021,049,050,055 |
| Unika | 051,052,041 |
| Universal | 051,052 |
| Viewstar | 007,008,019,021, 026,028,029,032,033,040,041 |
| Warner Amex | 044 |
| Zenith | 014,042,057 |

### VCR Codes

| Brand | Code |
|---|---|
| Aiwa | 015 |
| Akai | 003,017,022,023,065,066 |
| Audio Dynamics | 014,016 |
| Broksonic | 010 |
| Candle | 007,009, 013,044,045,046,052 |
| Canon | 008,053 |
| Capehart | 001 |
| Citizen | 007,009,013, 044,045,046,052 |
| Colortyme | 014 |
| Craig | 007,012 |
| Curtis Mathes | 000,007,008, 014,015,044,046,053,064,067 |
| Daewoo | 013,045,052 |
| DBX | 014,016 |
| Dynatech | 015 |
| Electrohome | 027 |
| Emerson | 008,009,010, 013,015,020,023, 027,034,041,042,047,049, 057,062,065,067,068,070 |
| Fisher | 002,012,018, 019,043,048,058 |
| Funai | 015 |
| GE | 000,007,008,052,053 |
| Goldstar | 009,014,046,060 |
| Harman Kardon | 014 |
| Hitachi | 005,015,035,036 |
| Instant Replay | 008 |
| JCL | 008 |
| JC Penney | 002,005,007,008, 014,016,030,035,051,053 |
| JVC | 002,014,016,030,046 |
| Kenwood | 002,014, 016,030,044,046 |
| Lloyd | 015 |
| Logik | 051 |
| Magnavox | 008,029,053,056 |
| Marantz | 002,008,014,016, 029,030,044,046,061 |
| Marta | 009 |
| MEI | 008 |
| Memorex | 008,009,012,015 |
| MGA | 004,027 |
| Midland | 032 |
| Minolta | 005,035 |
| Mitsubishi | 004,005, 027,035,040 |
| Montgomery Ward | 006 |
| MTC | 007,015 |
| Multitech | 007,015,031,032 |
| NEC | 002,014,016,030, 044,046,059,061,064 |

| Brand | Code |
|---|---|
| Panasonic | 008,053 |
| Pentax | 005,035,064 |
| Pentex Research + | 046 |
| Philco | 008,029,053,056 |
| Philips | 008,029 |
| Pioneer | 005,016,033,050 |
| Portland | 044,045,052 |
| ProScan | 000 |
| Quarta | 002 |
| Quasar | 008,053 |
| RCA | 000,005,007,008, 028,035,037,054,069 |
| Radio Shack/Realistic | 002, 006,009,009,012, 015,019,027,043,053 |
| Samsung | 007,013,022,032,042 |
| Sansui | 016,071 |
| Sanyo | 002,012 |
| Scott | 004,013,041,049,068 |
| Sears | 002,005,009,012, 018,019,035,043,048 |
| Sharp | 006,024,027,039,045 |
| Shintom | 017,026,031,055 |
| Sony | 017,026,038 |
| Sylvania | 008,015,029,053,056 |
| Symphonic | 015 |
| Tandy | 002,015 |
| Tashiko | 009 |
| Tatung | 030 |
| Teac | 015,030,069 |
| Technics | 008 |
| Teknika | 008,009,015,021 |
| Toshiba | 005,013,019,048,049 |
| Totevision | 007,009 |
| TMK | 067 |
| Unitech | 007 |
| Vector Research | 014,016,044 |
| Victor | 016 |
| Video Concepts | 014,016,044 |
| Videosonic | 007 |
| Wards | 005,006,007,008,009, 012,015,019,025,027,031,035 |
| Yamaha | 002,014,016,030,046 |
| Zenith | 011,017,026 |

### Laserdisc Player Codes

| Brand | Code |
|---|---|
| RCA | 033 |
| Pioneer | 033 |

1Q57 404-01A

**FIGURE 3-26**   A code listing for the Radio Shack 3-in-1 universal remote control.

Be sure to have the fresh batteries ready to install before you remove the old batteries. The universal remote's memory only lasts about a minute without the batteries in place. If the memory is lost, simply re-enter the 3-digit code for your remote control.

If the universal remote stops working after you successfully test the control of each device (or if you are unable to get the unit to work at all), make the following checks:

■ Press the device key for the electronic device that you want to control.
■ Replace the batteries.
■ Confirm that your remote controls are working properly using manual controls or the original remotes.
■ If some buttons do not function for your device, you might be able to scan to a better device code.

**Remote-control care and maintenance**   Your remote controls are an example of electronic devices of good design and workmanship, and should be treated with care. The following tips will help you enjoy these electronic wonders for many years.

■ Keep the remote unit dry. If a liquid spills into the unit, dry it off as soon as possible and dry it with a hair dryer. Liquids contain products that can corrode the electronic circuits.
■ Handle the remote control gently and carefully. Dropping it can damage its circuit boards and case and cause the control to work improperly.
■ Use and store the remote control only in normal room-temperature environments. Temperature extremes can shorten the life of electronic devices and distort or melt plastic parts.
■ Keep the remote control away from dust and dirt, which can cause premature wear of parts.
■ Wipe the remote control with a damp cloth occasionally to keep it looking new. Do not use harsh chemicals, cleaning solvents, or strong detergents to clean the remote control.
■ To prevent any internal damage, do not twist the remote unit.

**Remote-control extenders**   The remote-control extenders are radio-frequency (RF) devices that let you control TVs, lights, radios, and DBS satellite receivers from other rooms or even outside your home. Most remotes use infrared light signals and cannot go through walls of a building. Basically, they are a "line-of-sight" control device.

Now look at a remote-control extender made exclusively for the RCA DSS satellite dish receiver. This unit is made by Windmaster (904-892-7815) and is shown in Fig. 3-27. This remote extender will let you control the DSS receiver from any room in your home.

**Transmitter and receiver extender installation**   The transmitter attaches to the remote control and senses the infrared signals from the remote control. The transmitter converts the IR control signals into RF waves. Figure 3-28 shows the extender being installed.

The base receives the RF signal from the transmitter hand unit and converts these waves back to infrared signals.

Place the base receiver in front of your RCA DSS satellite receiver (as shown in Fig. 3-29). The base receiver must be located so that no obstructions will be between it and the receiver. Figure 3-30 shows the extender installed on the remote and ready for use.

**FIGURE 3-27** The Windmaster remote DBS extender control and receiver.

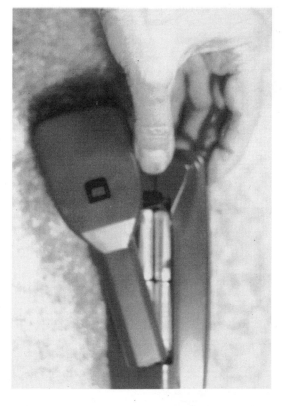

**FIGURE 3-28** The battery cover is removed from the DBS remote unit. Just snap the remote extender onto the RCA hand unit.

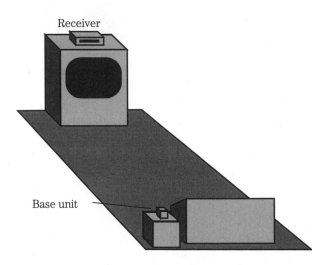

Receiver

Base unit

**FIGURE 3-29**    Placement and location of the remote extender base receiver.

**FIGURE 3-30**    The remote DSS Windmaster extender installed and being used.

### *What to do if you have trouble with the extender*

**1** Be sure that the base receiver is plugged into a working electrical ac outlet.

**2** If the red LED light inside the transmitter does not light when the remote control is operated, check the following:

■ Be sure that the battery inside the transmitter unit is good and installed correctly.

**3** If your receiver does not respond, but the red light inside the base receiver lights when the remote control is used, be sure that nothing is blocking its light path.

**4** If the red light on the transmitter lights when you operate the remote, but does not light in the base receiver, then check the following:

■ Move the antenna to a different position for better reception.
■ Lower the antenna rod.
■ Move the base receiver to another location and try to operate it again.

To order a DBS remote extender, call 1-800-624-4112.

# HOW COLOR
# TVs AND PC
# MONITORS WORKS

# The Color TV Signal

Before looking at the block diagram operation of a typical color TV, see what the color TV signal (which comes to your TV via antenna, cable or DBS dish) energy components contains.

The video signal that comes from the TV transmitter is an electrical form of energy that enters the free space in some form of electromagnetic waves. No one at this time has an explanation of what makes up this electromagnetic form of energy. It travels at the speed of light. For this transmitted wave to carry intelligence, it must be varied (modulated) in some way.

Your color TV is thus receiving from the TV transmitter visual images in full color, as well as audio sound. The U.S. color TV system must be compatible with black-and-white television standards. Compatibility is when the system produces programs in color on color TVs and black and white on monochrome TVs. Conversely, color TVs will receive B&W pictures when they are being transmitted.

The color TV signal contains two main components, luminance (black & white or brightness) information, and chrominance (color) information, which is added to the luminance signal within the color receiver circuits to produce full-color pictures.

The transmitted color TV signal contains all of the information required to accurately reproduce a full-color picture. The U.S. standard-frequency TV channel width for a color picture is 6 MHz.

The color signal contains not only picture detail information, but equalizing pulses, horizontal sync pulses, vertical sync pulses, 3.58-MHz color-burst pulses, blanking pulses, and VITS and VIRS test pulses. The horizontal blanking pulse is used to turn off the electron scanning beam from the gun in the picture tube, at the end of each scan sweep line. Vertical blanking pulses are used to blank out the beam at the top and bottom of the picture, so you will not see any of the transmitted pulses used for picture control and testing.

The color (chrome) is phase-and-amplitude encoded relative to the 3.58-MHz color burst and is superimposed on the black-and-white signal level. This video information is used to control the three electron beams (red, green, and blue) that scans from left to right with 525 horizontal lines across the face of the picture tube. How the color picture tube works is explained later in this chapter.

## COLOR TV SIGNAL STANDARDS

The signals from the color TV transmitter are reproduced on the screen of the TV to closely match those of the original scene. The black-and-white and color signals have an FM (frequency modulated) sound carrier and an AM (amplitude modulated) video carrier with a channel bandwidth of 6 MHz.

The portion of the black-and-white signal that carries the video picture information is called the *carrier amplitude*, which is the brightness or darkness of the original picture information modulation. Now, a portion of the color video signal that carries picture information consists of a composite of color information and amplitude variations. A unique feature is the 3.58-MHz color syncburst, which is located right back of the horizontal sync pulse. This color pulse is often referred to as "setting on the back porch of the horizontal sync pulse." Every horizontal scan line of video color information contains the picture information, horizontal blanking, sync pulses, and a color burst of at least eight cycles.

The picture information in a color TV signal is then obtained from the red, green, and blue video signals that the color TV camera generates as it scans the scene to be transmitted. The luminance (Y) and chrominance (C) signals derived from the basic red, green, and blue video color signals have all of the picture information required for transmission. They are then combined into a single signal by algebraic addition. The final color product contains a chrominance signal, which provides the color variations for the picture, and the luminance signal, which provides the variations in intensity or the brightness of the colors.

The three different color TV broadcast standards in the world are not compatible with each other. In the United States, the NTSC standard was devised by the National Television System Committee and in 1953, was approved by the FCC (Federal Communications Committee). The NTSC system is now used in the USA, Canada, Japan, and other countries. Most of Europe uses the PAL system, and the SECAM standard is used in Russia. Before the year 2006, all TV video signal transmissions will be changed over gradually to compressed digital video and sound format. This system was approved by the FCC in 1997.

# Color TV Receiver Operation

Now, follow a block diagram of a typical modern color TV receiver and see how the various sections work together to produce a color picture.

## THE "HEAD END" OR TUNER SECTION

The color TV signal enters the TV receiver at the electronic tuner. This is where you change stations on your TV, via an antenna, cable system, or a DBS satellite dish. Referring to the overall TV block diagram in (Fig. 4-1), you will note the electronic tuner has RF amplifier, oscillator, and mixer stages that convert the RF carrier signal to a 45.75-MHz signal IF that is fed to the video IF stages or strip. Most modern TV tuners are remotely controlled and have AFC (Automatic frequency control) to lock in the TV. Figure 4-2 shows the electronic tuner mounted in a late-model color TV.

## THE IF STAGES AND VIDEO AMPLIFIER/DETECTORS

After amplification and signal processing in the IF stages, the video and audio signals are found at the output of their detectors. The audio has now been detected in the 4.5-MHz audio IF section and stereo audio is fed to the right and left audio power amplifiers, then onto the speakers. Some TVs also have MTS/SAP and DBX decoders.

## VIDEO DETECTOR

An AM (amplitude modulation) detector, also called a *diode detector*, converts the picture frequency within the video IF stages down to a video frequency. Right after this detector is the 4.5-MHz sound trap to remove a frequency that would result in a heterodyning of the picture and audio frequencies of the video IF circuits. Without this trap, dark bars would appear across the picture and cause hum in the audio.

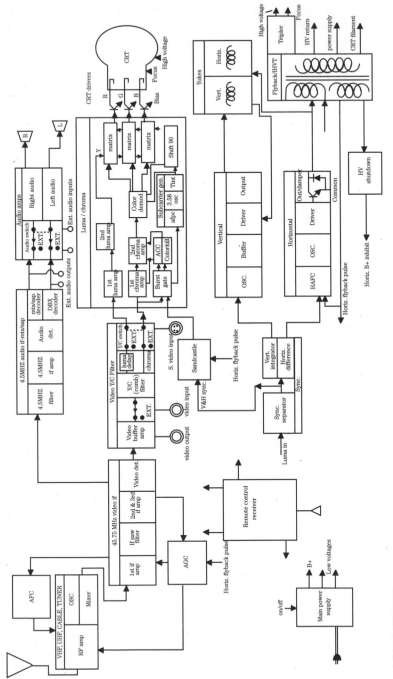

**FIGURE 4-1    A block diagram of color TV/monitor system.**

**FIGURE 4-2**    The electronic tuner "head end" used in a TV.

## VIDEO AMPLIFIERS

The video amplifiers amplify the video from the detector to levels great enough to drive other circuits.

## LUMA DELAY LINE

This is a device that electronically delays the Y signal (monochrome video), or luminance signal about 0.8 microseconds to match the delay in the chrome circuits. The reason for this is that the color signal is delayed more than the luminance because more circuits are used to process the color signal and they are tuned to a narrower-frequency bandwidth.

## THE CHROMA PROCESSING CIRCUITS

From the video detector, the signal goes to some video-processing stages. You will find a comb filter, used to produce sharper pictures and the delay line. Chroma and video signals along with vertical and horizontal sync signals are also obtained from the video stages.

## THE CHROMA AND LUMINANCE STAGES

The composite video signal from the detector is fed to the first chroma amplifier and also to the luminance (B&W) amplifier in the luma/chroma block. The 3.58-MHz oscillator and subcarrier generator is used to extract the chrominance signals. This is accomplished in the color demodulator stage. These signals go to the matrix circuits and then onto the cathodes of the color (CRT) picture tube guns. This block also contains blanking circuits, burst gate, and color killer stages.

## COLOR-KILLER CIRCUIT OPERATION

The color-killer circuit has a couple of basic functions. Its performance during black-and-white picture transmission is to keep high-frequency signals or noise from being amplified via the chroma amplifiers. This keeps the color snow or confetti from being seen within the picture. It also keeps you from seeing color rainbows from around fine detail and edges of a black-and-white picture. This stage is also used to kill the color signal during weak or snowy signal conditions. Thus, the killer circuit must know the difference between the color burst signal and interference or noise conditions.

## SANDCASTLE CIRCUIT OPERATION

Most modern color TVs have a sandcastle circuit located within an IC. The sandcastle circuit is a special device used by design engineers to inject three mixed signals into one pin of this IC. The IC separates these three signals and uses them for various internal functions. The three signals are the horizontal sweep pulses, a delayed horizontal sync pulse, and a vertical sweep pulse.

After separation inside the IC, the horizontal sweep, also called *flyback pulses*, provides horizontal blanking for the output signals of the chip, and the delayed sync pulses separates the color burst from the back porch of the horizontal sync pulse, and the vertical pulse provides vertical blanking. If you would look at this pulse on an oscilloscope that is developed from this chip, it appears as a sandcastle, thus the name given this circuit. If one of the input pulses is missing or the IC is defective, the TV screen will be blanked out.

## FUNCTIONS OF THE SYNC CIRCUITS

The basic function of the TV sync circuits is to separate the horizontal and vertical sync pulses from the video signal. These separated pulses are then fed to the horizontal and vertical sweep stages to control and lock-in the color picture. These circuits need good noise immunity to maintain good, stable vertical and horizontal picture lock in.

Some color sets have the sync and AGC (automatic gain control) circuits combined. Normally the AGC circuit develops a bias in proportion to the sync pulses peak-to-peak level, which is then used as a dc voltage to control TV receiver gain. Keyed AGC circuits are used because they provide better noise immunity.

## VERTICAL SWEEP DEFLECTION OPERATION

The vertical sweep oscillator stage receives a sync pulse from the vertical integrator stage, which forms and develops this pulse. This sync pulse keeps the vertical oscillator running at the vertical scanning rate. Some sets have digital countdown and divider circuits to perform this task. The oscillator feeds the buffer and driver stages. The output current from the vertical power amplifier stage is then applied to the vertical winding of the deflection yoke, which is located around the neck of the picture tube. A pulse from the vertical output stage is used for picture tube screen blanking. Some of these pulses can also be used for convergence of the three color beams in the gun of the picture tube.

## HORIZONTAL SWEEP DEFLECTION OPERATION

Older color TVs have a horizontal circuit that consists of a sawtooth generator that would drive the horizontal sweep and high-voltage generating transformer. This circuit is controlled by an AFC (automatic frequency control) circuit that compares the frequency of the oscillator with the sync pulse coming from the TV transmitter and then produces a correction dc voltage for any oscillator frequency drift that might occur. The deflection current for the horizontal yoke coils, along with picture tube high voltage and focus voltage, plus other pulses, are generated by the horizontal sweep transformer. The horizontal sweep and HV stages need very good voltage regulation to produce a good sharp color picture.

Figure 4-3 shows various adjustment controls found on most TVs. These are horizontal hold (Horiz. Hld.), brightness level, vertical hold (Vert. Hld.), vertical height/vertical linearity, and sometimes a color killer and AGC adjustment.

Modern color TVs use a digital countdown divider system to generate and control pulses to drive the horizontal sweep stage. Figure 4-4 shows the horizontal sweep and

**FIGURE 4-3**    **The adjustment control location on back of a TV.**

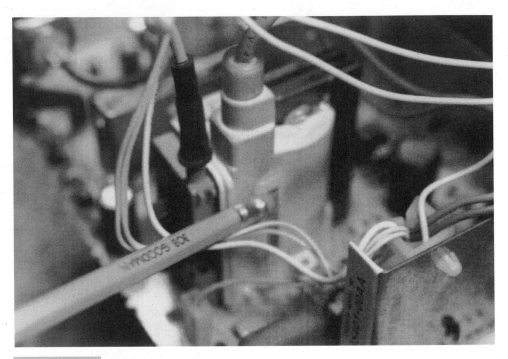

**FIGURE 4-4** A TV's high-voltage transformer.

high-voltage transformer section. Figure 4-5 shows the high-voltage lead and cup that plugs into the picture tube HV button. This lead will carry a voltage of 25,000 to 32,000 volts. Use caution because this voltage can still be present at this cap—even when the TV is turned off.

The horizontal output stage in these modern TVs are usually of a pulse-width design that not only sweeps the electron beam across the picture tube, but also develops other dc voltages to operate other circuits in the color TV chassis. This eliminates the heavy weight and costly price tag of a power transformer and also improves the efficiency of the ac power and current that the TV uses. A safety high voltage shutdown circuit is also used.

## SOUND CONVERTER STAGE OPERATION

This stage converts the audio frequency in the video IF passband circuit (41.25 MHz) down to the sound IF frequency of (4.5 MHz) by heterodyning (beating) the picture and sound frequencies together and tuning to the different frequency. This stage is usually a diode detector and a 4.5-MHz tuned circuit.

## SOUND IF AMPLIFIER OPERATION

This stage is sometimes called an audio IF amplifier. This stage is used for amplifying the 4.5-MHz sound IF signal to a level that is high enough to be detected by the sound detector.

## THE SOUND (AUDIO) DETECTOR

The sound detector is also called the *audio detector*. This part of the circuit converts the FM-modulated 4.5-MHz sound IF frequency to an audible sound frequency that drives an audio amplifier stage(s).

## AUDIO AMPLIFIER STAGE

This circuit is used to amplify the audio frequencies from the detector to power levels great enough to drive the speaker(s). If the TV receiver is MTS equipped, there might be two amplifiers (stereo) and the MTS-decoding circuits will precede the audio amplifiers.

## TV POWER-SUPPLY OPERATION

As with any electronic device, the power supply is the heart of the device and makes all the other systems operate. If the TV is dead or not working properly, look at and check out the power-supply section first. The problem could be simple, such as an off/on switch that is defective, a blown fuse, or tripped "off" ac breaker for the wall socket. And check the condition of the power cord and be sure that it is plugged into the wall socket. Figure 4-6 shows the location of the main TVs power supply fuse located on the circuit board. Replace fuse with the same current rating if it has blown. If it blows again, suspect a short circuit or power-supply problems.

**FIGURE 4-5**    Location of the picture tube's HV anode connection. Use caution in this area of the TV.

FIGURE 4-6   **The main power clip-in fuse in a TV.**

The transistor power supply regulator heatsink is shown in Fig. 4-7. If the regulator or sweep output transistors are defective (shorted), they might cause the main fuse in the power supply to blow.

This should now give you an overview of how these blocks and circuits in a color TV work together and what could go wrong. Take a closer look at the circuit operation within some of these blocks and see how they work and what to do when they don't.

## SWEEP CIRCUITS AND PICTURE TUBE OPERATIONS

The following section of this chapter shows some advanced troubleshooting of color TV and computer monitor circuits. The horizontal and vertical sweep circuit operations and troubleshooting will be all worked in together.

**Vertical sweep circuit operation**   You can use the following information for color TV and computer monitor troubleshooting to isolate the vertical oscillator, driver, and sweep output stage problems. The vertical driver and output stages amplify the vertical oscillator signal, which provides the current drive needed for the vertical deflection of the yoke. A defective driver circuit, output stage, or yoke can cause loss of deflection, reduced height, or vertical linearity picture problems. Before you use signal injection to troubleshoot a vertical sweep problem, use a dc voltmeter (DVM) to confirm that you have proper bias voltages on the output stage components. The vertical stages are usually dc coupled to get good linearity. A wrong dc voltage affects all the components in the oscillator, driver, and output stages. A dc bias problem must be repaired before you can effectively use signal injection in the vertical stages. Use an analyzer, such as the Sencore VA62 or TVA92, to inject vertical and horizontal sweep signals into the circuit (Fig. 4-8).

**Collapsed vertical raster**   This problem will show up as a thin white horizontal line across the screen (Fig. 4-9). To isolate the trouble, inject the analyzer's vertical drive signal into the output of the vertical driver circuit (Fig. 4-10).

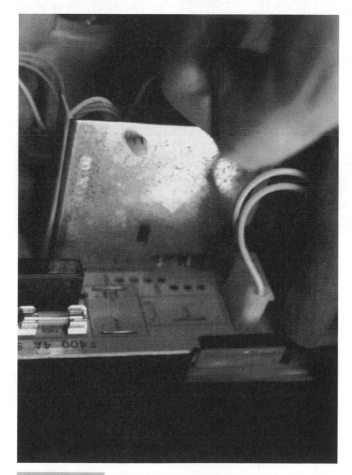

**FIGURE 4-7**   The heatsink for the regulator transistor and fuse that could blow if transistor becomes shorted.

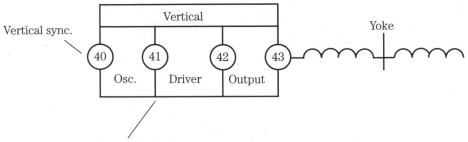

**FIGURE 4-8**    This block diagram shows where to inject the vertical sync test signal.

**FIGURE 4-9**    A thin, white horizontal line indicates a vertical sweep failure.

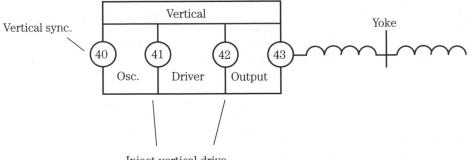

**FIGURE 4-10**    Arrows indicate where to inject the vertical drive test signals.

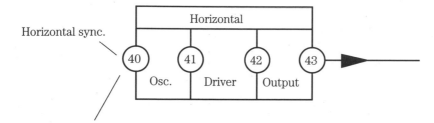

Inject horizontal sync.

**FIGURE 4-11**    **The signal is being injected into the horizontal oscillator stage.**

Injecting a test signal into the vertical stages will not always produce full vertical picture deflection because most of the signals are uniquely shaped by feedback loops and waveshaping circuits.

Look for the sweep (raster) to expand on the screen, but remember that it will probably not be a full screen. If the raster expands, either partially or fully, the circuits from the injection point to the output stage are good. If the sweep does not expand, check the output components or "ring test" the deflection yoke coils.

The vertical drive test signal will not directly drive the vertical yoke coils.

**Isolating horizontal sync problems**    The horizontal sync pulses control the timing of the horizontal oscillator. Many computer monitors receive horizontal sync directly. TV receivers and some monitors have a composite sync, or "sync on video" input that requires the use of sync separators. Sync pulses that are low in amplitude, the wrong frequency, or are missing will cause the monitor to lose horizontal picture hold.

**Loss of horizontal sync symptoms**    Inject the video analyzer's horizontal sync drive signal into the input of the horizontal oscillator (Fig. 4-11). If the TV or monitor regains horizontal hold control and produces full horizontal deflection, the driver and output stages are operating properly. The next step is to troubleshoot the horizontal sync circuit path. If the TV or monitor displays the same symptoms with the drive signal from the analyzer applied, then troubleshoot the horizontal oscillator circuit.

## DEFLECTION YOKE PROBLEMS

The changing current through the windings of the deflection yoke produces a magnetic field that scans the electron beam across the face of the picture tube. Yokes can develop shorted or open windings. An open or shorted winding might cause reduced vertical or horizontal size, or a complete loss of deflection.

The analyzer ringer test can be used to find defective yokes—even if it has a single shorted turn. Readings of 10 rings or more are accompanied by a "good" display and shows that the winding does not have a shorted turn. "Bad" readings, less than 10 rings, indicate a shorted turn(S).

**Collapsed raster symptom**   For this symptom, picture would be pulled in and small; you need to ring the horizontal and vertical yoke windings. For this test, always disconnect the yoke from the circuit and unsolder any damping resistors (leave the yoke mounted on the picture neck).

If the horizontal and vertical yoke windings ring more than 10 rings, the yoke is good. If any of the windings ring less than 10, the yoke is defective and needs to be replaced.

## KEY VOLTAGE READINGS

A lot of troubleshooting information can be revealed about a TV's operation by measuring the dc and peak-to-peak voltage at the collector of the horizontal output transistor. The Sencore analyzer has a dc and peak-to-peak voltmeter with the input protection required for measuring signals at this test point. The dc reading shows you if the B+ supply is working correctly, while the peak-to-peak reading shows if the output circuits are developing the needed high voltage.

## INOPERATIVE COMPUTER MONITOR PROBLEM

With the analyzer or voltmeter, measure the dc voltage at the collector of the horizontal output transistor. If the B+ voltage is low or missing, unload the power supply by disconnecting the collector of the horizontal output transistor from the circuit. Measure the voltage at the output of the power supply regulator again. If the voltage is low or missing, troubleshoot the power supply. If the voltage is lower than that noted on the schematic, then something is loading down the supply. In this case, troubleshoot the output transistor, flyback transformer, or yoke (Fig. 4-12).

**Testing sweep high-voltage transformers**   The sweep or flyback transformer in a TV or computer monitor is used to develop the focus, high voltage, and other scan-derived power-supply voltages. The flyback is a high-failure component and it is also one of the most expensive parts in the TV or computer monitor.

Although an open transformer winding is easy to identify using an ohmmeter; the more common shorted transformer winding is nearly impossible to locate using the conventional testing methods. The Sencore analyzer has a patented ringer test that provides you with an easy-to-use, fail-safe method of finding opens and shorts in high-voltage sweep transformers.

## ANOTHER MONITOR PROBLEM

For this test, connect the ringer across the flyback's primary winding and ring test the transformer. A "good" reading of 10 rings or more indicates that none of the windings in the flyback have shorts or opens. You do not need to ring any other winding. A shorted turn in any other winding will cause the primary to ring bad (Fig. 4-13).

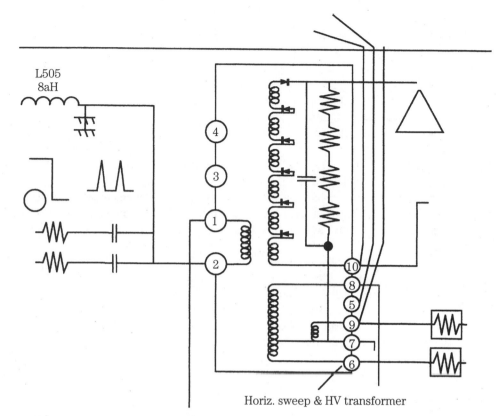

**FIGURE 4-12**    The Sencore's OVM meter has input protection to measure the high P-P HV pulses in this horizontal sweep and HV circuit.

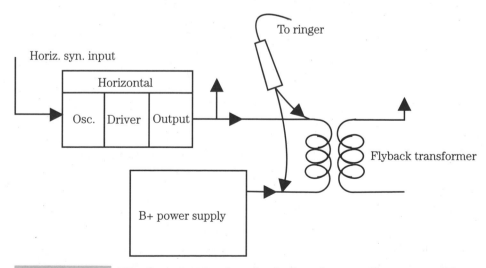

**FIGURE 4-13**    The ringer test hook-up for finding shorts and opens conditions in a flyback transformer.

A "bad" reading, less than "10" rings, may be caused by a circuit connected to the flyback that is loading down the ringer test. Disconnect the most likely circuits in the following order:

**1** Deflection yoke.
**2** CRT (picture tube). Unplug the socket connection.
**3** Horizontal output transistor collector.
**4** Scan-derived supplies.

Retest the flyback after you have disconnected each circuit. If the flyback now rings "good," it does not have a shorted winding.

If the flyback still checks out bad after you have disconnected each circuit, unsolder it and completely remove it from the circuit. If the flyback primary still rings less than 10, the flyback is defective and must be replaced.

**Testing the high-voltage diode multipliers**  During normal TV/monitor operation, a large pulse appears at the collector of the horizontal output transistor. The output connects to the primary of the flyback transformer and the pulses are induced into the flyback's secondary. The pulses are stepped up and rectified to produce the focus and high voltages. These voltage pulses are rectified by high-voltage diodes contained in the flyback package or in an outboard diode multiplier package.

Because these are high-voltage components, it is often difficult to determine dynamically if the diodes will break down under high-voltage conditions. The Sencore analyzer has a special test to determine if these diodes are good or bad.

## HIGH-VOLTAGE PROBLEMS

It is only necessary to do this test if all of the following conditions are met:

**1** The high voltage or focus voltage is low or missing.
**2** The B+ and peak-to-peak voltages at the horizontal transistors are normal.
**3** The horizontal sweep (flyback) transformer passes the ringer test.

With the analyzer, feed a 25-volt peak-to-peak horizontal sync drive signal into the primary winding of the flyback transformer. The step-up section of the transformer and the high-voltage diodes should develop a dc voltage between the second anode and high-voltage resupply pin on the flyback transformer. Measure this voltage with a dc voltmeter. Look up this voltage on the schematic to determine if the high-voltage diodes are good or bad.

## HORIZONTAL OSCILLATOR, DRIVER AND OUTPUT PROBLEMS

If the horizontal yoke, flyback, multiplier, horizontal output transistor, and B+ supply have tested good, but the TV still lacks deflection or high voltage, the horizontal driver circuit might be defective. A missing or reduced-amplitude horizontal drive signal could prevent the TV

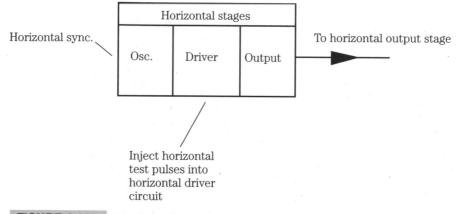

**FIGURE 4-14** **The injection point for test pulses into the horizontal driver circuit.**

from starting and operating properly. Use the Sencore analyzer's horizontal drive signal to isolate problems in the horizontal drive circuit. Refer to the signal injection points in Fig. 4-14.

**TV start-up problem**

**1** Before injecting into the horizontal drive circuit, test the flyback and yoke, the high-voltage multiplier, the horizontal output transistor, and the 13-V supply.
**2** When injecting at the output transistor, disconnect the secondary winding of the driver transformer from the base.

Inject the horizontal drive signal into the driver circuit. Watch for horizontal deflection on the picture tube. If it returns, you are injecting after the defective stage. If nothing happens, inject the horizontal drive signal at the base of the horizontal output transistor. Refer to Fig. 4-15 for these injection points.

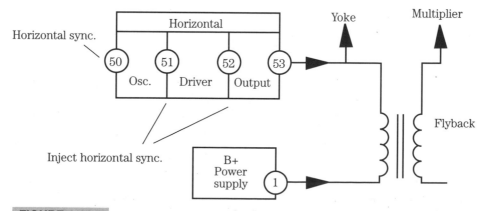

**FIGURE 4-15** **Horizontal drive signal test injection points. The base of the output is a good injection point.**

**How to measure the TV's high voltage**   The picture tube (CRT) requires a very high dc voltage to accelerate the electrons toward the screen of the CRT. This voltage is developed in the secondary winding of the flyback transformer and is amplified and rectified by the integrated diodes in the flyback, or by a separate multiplier circuit.

Measuring the high voltage at the second anode of the picture tube lets you know if the output circuit, sweep transformer, high-voltage multiplier, and power-supply regulation circuits are working properly. Additionally, some TVs and computer monitors have adjustments for setting the high voltage and focus voltage correctly.

> Use extreme care when measuring and adjusting any voltage around the picture tube and high-voltage power supplies.

**Blurred, out-of-focus picture symptom**   For this problem, first measure the picture-tube high-voltage capacitor with a high-voltage probe. Compare these readings with the HV readings shown in the schematic. Also, if these voltages are OK, check the focus voltage and suspect that the CRT is weak.

**Switching transformer checks**   Switching transformers are used in power-supply circuits to step voltages up or down. They are one of the most common components to fail in switch-mode power supplies. Open windings are easy to find with an ohmmeter, but shorted turns are nearly impossible to locate with conventional test methods. The Sencore video analyzer's ringer test helps you to locate switching transformer with open or shorted windings.

> For this test, the switching transformer must be removed from the TV's circuit.

To perform this test, connect the Sencore analyzer ringer test leads across a winding on the switching transformer. A reading of 10 rings or more will show that the winding does not have a shorted turn. Perform this same test on all windings of the switching transformer.

## THE VERTICAL SWEEP SYSTEM

In my feedback from many electronic technicians, most say that the vertical sweep systems are among the most difficult circuits in a monitor or TV to troubleshoot. Even the most small change in a component can cause reduced sweep deflection, nonlinear deflection, or picture fold-overs. These symptoms can be caused by a small circuit part or an expensive vertical yoke. Thus, you must think carefully of a strategy to take the guesswork out of isolating vertical sweep problems.

**How vertical deflection works**   Knowing how the vertical sweep deflection circuits operate requires an understanding of picture tube beam deflection. The electron beam travels to the face of the picture tube striking the phosphor surface coating to produce light on the front of the picture tube.

If the stream of electrons travels to the face plate of the tube without any control from any magnetic or electrostatic field, the electrons will strike the center of the screen and produce a white dot. To move this dot across the face of the picture tube screen requires that the electrons be influenced by an electrostatic or magnetic field.

In video display tubes, a magnetic field is produced by the vertical coils of a yoke mounted around the neck of the tube. The yoke is constructed with coils wound around a magnetic core material.

When current flows in the vertical yoke coil windings, a magnetic field is produced. The yoke's core concentrates the magnetic field inward through the neck of the picture tube. As the electrons pass through the magnetic field on the way to the tube's face plate, they are *deflected* (pulled upward or downward) by the yoke's changing magnetic field. This causes the electron stream to strike the picture tube face plate at points above and below the screen center.

To understand how electrons are deflected requires a review of the interaction of magnetic fields. As you refer to Fig. 4-16A and 4-16B, you might recall that an individual electron in motion is surrounded by a magnetic field. The magnetic field is in a circular motion surrounding the electron. As electrons travel through the magnetic field of the yoke, the magnetic fields interact. Magnetic lines of force in the same direction create a stronger field, but magnetic lines in opposite directions produce a weaker field. The electrons are then pulled toward the weaker field.

The direction of the current in the yoke coil determines the polarity of the yoke's magnetic field. This determines if the electron beam is deflected upward or downward.

How far the electrons are repelled when passing through the yoke's magnetic field is determined by the design of the yoke and the level of current flowing through the vertical coils. The higher the current, the stronger the magnetic field and resulting electron deflection.

A requirement of vertical sweep deflection in a TV or monitor is that the current in the coils of the vertical yoke increase an equal amount for specific time intervals. This linear

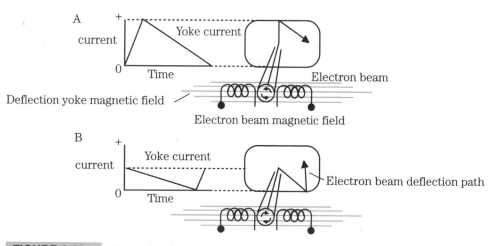

**FIGURE 4-16**    **The yoke mounted on the CRT neck produces the magnetic field, resulting in electron deflection.**

current change causes the deflection of the electron beam from the top to the center of the picture tube faceplate.

The waveforms shown in Fig. 4-16A and 4-16B represents a current increasing and decreasing in level, with respect to time. Figure 4-16A shows the current increasing quickly and then decreasing slowly back to zero. This would cause the electron beam to quickly jump to the top of the picture tube screen and then slowly drop back to the center.

Figure 4-16B shows the current increasing slowly in the opposite direction and then decreasing quickly back to zero. This would cause the electron beam to slowly move from the center to the bottom of the picture tube faceplate and then return quickly to the center.

During normal TV or monitor operation, the yoke current increases and decreases (Fig. 4-16A and 4-16B). The current changes directions alternating between the illustrations at approximately 60 times per second. The alternating current moves the electron beam from the top of the picture tube faceplate to the bottom and quickly back to the screen's uppermost area.

**How the vertical drive signal is developed**   The vertical circuit stages of the TV are responsible for developing the vertical drive signals. This signal is fed to an output amplifier, which produces alternating current in the vertical deflection yoke.

The vertical section consists of four basic circuits or blocks (Fig. 4-17). These include:

**1** Oscillator or digital divider.
**2** Buffer/pre-driver amplifier.
**3** Driver amplifier.
**4** Output amplifier.

The circuitry for these stages can be discrete components on the circuit board or might be included as part of one or more integrated circuits.

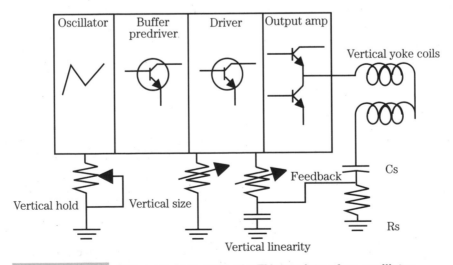

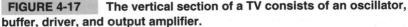

**FIGURE 4-17**   **The vertical section of a TV consists of an oscillator, buffer, driver, and output amplifier.**

The vertical oscillator generates the vertical sweep signal. This signal is then fed to the amplifiers and drives the yoke to produce deflection. Vertical oscillators can be free running or the more modern digital divider generators.

These free-running oscillators use an amplifier with regenerative feedback to self generate a signal. More common types are RC (resistance-capacitance) oscillators associated with ICs or discrete multivibrator or blocking oscillator circuits.

A digital divider generator uses a crystal oscillator. The crystal produces a stable frequency at a multiple of the vertical frequency. Digital divider stages divide the signal down to the vertical frequency. You will usually find most of the digital divider oscillator circuitry located inside an integrated circuit.

The output of a vertical oscillator must be a sawtooth-shaped waveform. A ramp generator is often used to shape the output waveform of a free-running oscillator or digital divider. A ramp generator switches a transistor off and on, alternately charging and discharging a capacitor. When the transistor is off, the capacitor charges to the supply voltage via a resistor. When the transistor is switched on, the capacitor is discharged.

The vertical oscillator must then be synchronized with the video signal so that a locked-in picture can be viewed on the picture tube. The oscillator frequency is controlled in two ways.

**1** A vertical hold control might be used to adjust the free-running oscillator close to the vertical frequency.
**2** Vertical sync pulses, removed from the video signal, are applied to the vertical oscillator, locking it into the proper frequency and phase.

If the oscillator does not receive a vertical sync pulse, the picture will roll vertically. The picture will roll upward when the oscillator frequency is too low and downward when the frequency is too high.

Several intermediate amplifier stages are between the output of the vertical oscillator and output amplifier stage. Some common stages are the buffer, predriver, and/or driver. The purpose of the buffer amplifier stage is to prevent loading of the oscillator, which could cause frequency instability or waveshape changes.

The predriver and/or driver stages shape and amplify the signal to provide sufficient base drive current to the output amplifier stage. Feedback maintains the proper dc bias and waveshape to ensure that the current drive to the yoke remains constant as components, temperature, and power-supply voltages drift. These stages are dc coupled and use ac and dc feedback, similar to audio amplifier stages.

Notice that ac feedback in most vertical circuits is obtained by a voltage waveform derived from a resistor placed in series with the yoke. The small resistor is typically placed from one side of the yoke to ground. A sawtooth waveform is developed across the resistor as the yoke current alternates through it. This resistor provides feedback to widen the frequency response, reduce distortion, and stabilize the output current drive to the yoke. This vertical stage feedback is often adjusted with gain or shaping controls, referred to as the *vertical height* or *size* and *vertical linearity controls*.

The dc feedback is used to stabilize the dc voltages in the vertical output amplifiers. The dc voltage from the output amplifier stage is used as feedback to an earlier amplifier stage. Any slight increase or decrease in the balance of the output amplifiers is offset by slightly changing the bias. Because the amplifier's waveforms are slightly distorted, the bias change will shift the bias on the output transistors, somewhat, thus bringing the stage back into compliance.

Much of the difficulty in troubleshooting vertical stages is caused by the feedback and dc coupling between stages. A problem in any amplifier stage, yoke, or its series components alters all of the waveforms and/or dc voltages, making it difficult to trace the problem.

**Vertical picture-tube scanning** The vertical output stage produces yoke current that then pulls the electron beam up and down the face of the picture tube. The vertical yoke might require up to 500 mA of alternating current to produce full picture tube deflection. A power output stage is now required to produce this level of current.

A current output stage commonly consists of a complementary symmetry circuit with two matched power transistors (Fig. 4-18). The transistors conduct alternately in a push-pull arrangement. The top transistor conducts to produce current in one direction to scan the top half of the picture. The bottom transistor conducts to produce current in the opposite direction to scan the bottom part of the picture.

Most vertical output stages are now part of an IC package and are powered with a single positive power supply voltage. The voltage is applied to the collector on the top transistor.

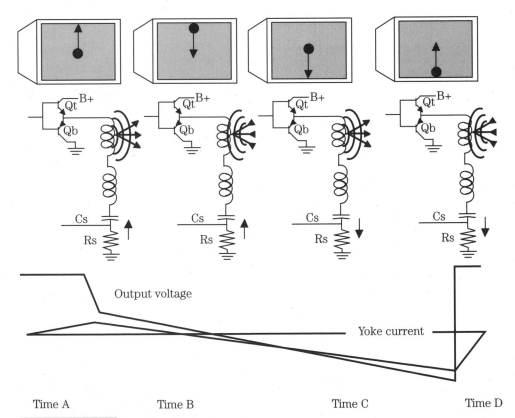

**FIGURE 4-18**    **The deflection currents and waveforms during four time periods of the vertical cycle.**

In this balanced arrangement, the emitter junction of the transistor should measure about one half of the supply voltage on this stage. In series with the vertical yoke coils is a large-value electrolytic capacitor. This capacitor passes the ac current to the yoke, but blocks dc current to maintain a balanced dc bias on the output amplifier transistors.

To better understand how a typical vertical output stage works, let's walk through the current paths at four points in time, during the vertical cycle illustrated in (Fig. 4-18). Starting with time A, the top transistor, Qt is turned on by the drive signal to its base. The transistor is biased on, resulting in a low conduction resistance from collector to emitter, which provides a high level of collector current. This puts a high plus (+) voltage potential at the top of the deflection yoke, resulting in a fast rising current in the yoke.

During time A, capacitor Cs charges toward a positive (+) voltage and current flows through the yoke and the top transistor, Qt. This pulls the picture tube's electron beam from the center of the picture tube up quickly to the top. During time A, an oscilloscope connected at the emitter junction displays a voltage peak, shown as the voltage output waveform in Fig. 4-18. The inductive voltage from the fast-changing current in the yoke and the retrace "speed-up" components cause the voltage peak to be higher than the positive (+) power supply voltage.

The current flowing in the deflection yoke during time A produces a waveform, as viewed from the bottom of the yoke to ground. This is the voltage drop across Rs, which is a reflection of the current flowing through the yoke.

During time B, the drive signal to Qt slowly increases the transistor's emitter-to-collector resistance. Current in the yoke steadily decreases as the emitter-to-collector (E-C) resistance increases and thus reduces the collector current. The voltage at the emitter junction falls during this time and capacitor Cs discharges. A decreasing current through the yoke causes the picture tube's electron beam to move from the top to the center of the screen.

To produce a linear fall in current through the yoke during time B demands a crucially shaped drive waveform to the base of Qt to meet its linear operating characteristics. The drive waveform must decrease the transistor's base current at a constant rate. Thus, the transistor must operate with linear base-to-collector current characteristics. These reductions in base current must result in proportional changes in collector current.

At the end of time B, transistor Qt's emitter-to-collector resistance is high and the transistor is approaching the same emitter-to-collector resistance as the bottom transistor, Qb. Capacitor Cs has been slowly discharging to the falling voltage at the emitter junction of the output transistors. Just as the voltage at the emitter junction is near one half of the positive (+) supply voltage, the bottom transistor begins to be biased ON to begin time C. This transition requires that the conduction of Qt and Qb at this point be balanced to eliminate any distortion at the center of the picture-tube screen.

During time "C", the resistance from the collector to emitter of transistor Qb is slowly decreasing because of the base drive signal and the increase of collector. The signal passes from capacitor Cs through the yoke and Qb. As Qb's resistance decreases and its collector current increases, the voltage at the emitter junction decreases. This can be seen on the voltage output waveform as it goes from one half positive (+) supply voltage toward ground during time C. The current increases at a linear rate through the yoke, as shown in the yoke current or voltage across Rs waveform (Fig. 4-18).

The resistance decrease of Qb must be the mirror opposite of transistor Qt's during time B. If not, the yoke current would be different in amplitude and/or rate, causing a difference in picture-tube beam deflection between the top trace and bottom trace times. At the end of time C, the emitter-to-collector resistance of Qb is low and Qb is slowly decreasing by the base increase of collector begins to discharge, producing current as the deflection yoke approaches a maximum level.

At the start of time D, the emitter-to-collector resistance of Qb is increased rapidly and collector current will decrease. This quickly slows the discharging current from capacitor Cs through the yoke and transistor. As the current is reduced, the trace is pulled quickly from the bottom of the screen back to its center. Time A begins again and the cycle is repeated again. This should now give you an overall view of how the horizontal and vertical sweep and scanning system produces a picture on your TV or computer monitor.

The basic inner workings of the color TV and PC monitor have now been covered. Another very important part of the color TV is the portion that you look at, the *picture tube* (*cathode-ray tube* or *CRT*).

**The working of the color picture tube**    The CRT works by producing (emitting) steady flow of electrons from the electron gun at the base (neck) of the CRT. These electrons are attracted to and strike the phosphor-coated screen of the CRT, causing the phosphors to emit light. Deflection circuits and a yoke outside the CRT produce a changing magnetic field that extends inside the CRT and deflects the beam of electrons to regularly scan across the entire face of the CRT, lighting the entire screen. The CRT can be divided into three functional parts (Fig. 4-19):

1 The electron gun cathode assembly.
2 The electron gun grids.
3 The phosphor screen and front plate.

The color picture tube is the last component in the video chain that lets you actually view a color picture on your TV or monitor. The major sections of a color set have previously been explained in this chapter, so now see how the CRT develops a color picture.

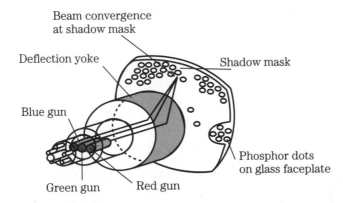

**FIGURE 4-19**    **An inside view drawing of a picture tube that has an in-line guns, metal dot mask, and phosphor dot triads on a glass faceplate.**

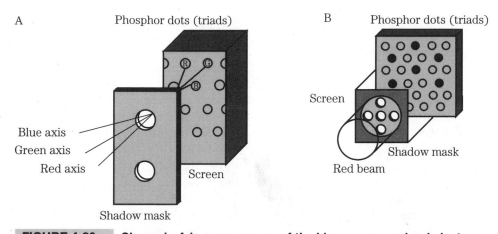

**FIGURE 4-20**    Shown in A is convergence of the blue, green, and red electron beams at the shadow mask. In B, each beam illuminates more than one phosphor triad.

Color CRTs use a metal shadow mask, phosphor screen, and three electron guns to produce red, blue, and green (RBG) colors that can produce a full color picture. These three colors are produced from phosphors that are excited by electron beams coming from three different guns, one gun for each of the (RBG) colors.

Figure 4-19 shows the relationship between shadow mask, electron guns, and the phosphors on the tube's faceplate. As you refer to Fig. 4-20A, notice that each beam converges through a hole in the shadow mask, while approaching the hole at a slightly different angle. Because of these different angles, the red beam hits the red phosphor, the blue beam the blue phosphor, and the green beam hits the green phosphor. However, each beam strikes more than one hole (Fig. 4-20B). With signals from the TV's red, green, and blue demodulators, these three electron beams are then mixed (matrixes) to different proportions to produce a very wide range of spectrum colors and intensities.

Over the years, TVs and monitors have used various types of picture-tube construction. The first color picture tubes used a delta gun arrangement with a dot shadow mask. As shown in Fig. 4-20A and 4-20B, the metal mask has evenly spaced holes with RGB phosphors clustered on the glass faceplate in groups of three. However, this triad arrangement had convergence problems because the three beams could not be made to meet at the shadow mask holes for certain areas of the faceplate.

**How the electron CRT gun works**    The electron gun consists of several different parts that together create, form, and control the electron beam. These parts are the filament (heater), cathode (K), the screen grid (G1), and the screen grid (G2). A monochrome (single color) CRT has just one electron gun, and a color CRT has three separate electron guns—one each for each color: red, green, and blue.

The cathode (K) is the source of the electrons, which are attracted to the screen. The cathode in most picture tubes look like a tiny tin can with one end cut out. It is coated with a material (such as barium or thorium) that emits large numbers of electrons when heated to a high temperature with the filament.

Several grids in front of the cathode attract the electrons away from the cathode toward the phosphor screen, control the rate of electron flow, and shape the cloud of electrons into a sharply focused beam.

The filament is mounted inside the cathode, and resembles the filament in a light bulb. The filament is used to heat the cathode. The filament is also called the *tube heater*. The filament is insulated from the cathode and does not make electrical contact.

The control grid is used to control the electrons. Without the control grid, the electrons would quickly leave in one big cloud with no control. The operation of the control grid can be compared to how a water faucet controls the flow of water.

***GE in-line electron gun***    General Electric developed the in-line gun with the slotted shadow mask in the mid 1970s. The metal mask has vertical slots instead of holes and the phosphors on the glass faceplate are RGB vertical strips, instead of dot triodes. The advantage of the in-line gun (Fig. 4-21) is simplification of convergence adjustments and a brighter picture level. When mated with properly designed yokes, the color convergence is considerably simplified. The Trinitron picture tube, invented and developed by the Sony Corporation, has a similar in-line design, except it has a three-beam electron gun and the shadow mask has a series of strips. The three common CRT gun patterns are in-line, delta, and Trinitron (Fig. 4-22).

In most cases, TV images are usually blobs of intensity and color. When a camera pans from one object to another, they are fuzzy because of bandwidth limitations of the video signal. In most cases, the images on computer displays consist of lines with sharp transi-

**FIGURE 4-21**    **This photo shows the GE in-gun assembly and the adjustments used for convergence.**

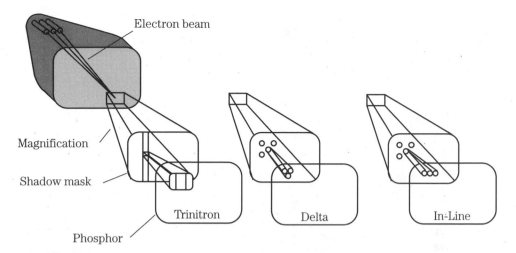

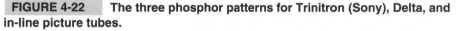

**FIGURE 4-22**    **The three phosphor patterns for Trinitron (Sony), Delta, and in-line picture tubes.**

tions of luminance. Usually the in-line/strip and slotted-mask CRT provides excellent pictures, but the in-line gun/dot mask-construction design displays text and graphics much better. A PC monitor color picture tube gun and socket assembly is shown in Fig. 4-23.

***Color picture tube summation***    A color TV contains all of the circuitry of the monochrome receiver, plus the added circuits needed to demodulate and display the color portion of the picture. To display the picture in color, three video signals are derived: the original red, green, and blue video signals.

The color CRT contains three color phosphors, each of which glows with one of the three primary colors when bombarded by electrons. These phosphors are placed on the inner surface of the picture tube faceplate as either triangular groups of the three colors (used in older models), alternating rectangles of the three colors, or alternating stripes of the three colors. Regardless of the version, all color tubes require three separate electron beams, each modulated with the video of one of the primary colors. All color tubes have some type of shadow mask placed behind the phosphors. This mask has a series of openings that allow each electron beam to strike only the correct color of phosphor.

The three beams must be precisely aligned to enable them to enter the opening in the mask at the correct angle and strike the correct phosphor. Stray magnetic fields could create enough error to cause the incorrect color to be displayed in parts of the picture. For this reason, color TVs have a coil mounted around the CRT faceplate and an automatic degaussing circuit to keep the picture tube and other nearby metal parts demagnetized.

Sometimes when a TV is moved to another location, the picture tube might have to be manually degaussed to clean up the color picture (Fig. 4-24).

## LARGE-SCREEN PROJECTION TV OPERATION

Large-screen projection TVs are now produced in many screen sizes and price ranges. Most have provisions for "surround-sound" audio amplifier systems, audio/video, and cable TV

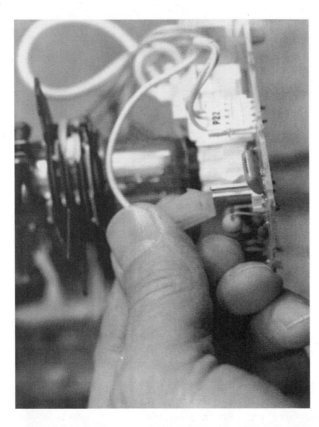

**FIGURE 4-23**    The picture tube socket and PC board assembly.

**FIGURE 4-24**    A degaussing coil being used to demagnetize a color TV picture tube.

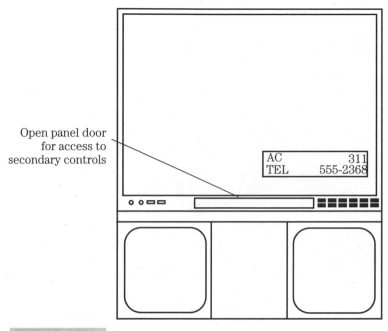

Open panel door for access to secondary controls

AC            311
TEL      555-2368

**FIGURE 4-25**    The front view of typical projection color TV.

and DBS dish input connections. A front view of a typical large-screen projection set is shown in Fig. 4-25. This type of TV projects the picture image onto the back of a translucent (Fresnel) screen that can then be viewed from the front. As shown in Fig. 4-26, the inside view these sets have three separate red, green, and blue (RGB) projection tubes to produce a bright picture.

A front-screen projection TV is illustrated in Fig. 4-27. These sets also use three separate red, green, and blue tubes to throw an image on a beaded projection screen, usually mounted on a wall.

In large-screen projection sets, high-definition, liquid-cooled projection tubes are used to provide a bright, high-resolution, self-converged picture display. Optical coupling is used between the projection tubes and the projection optics for display contrast enhancement. A screen with high-gain contrast and an extended viewer angle are now used on the newer-model projection receivers. Also, fault-mode sensing and electronic shutdown circuits are provided to protect the TV in the event of a circuit fault mode or picture tube arc.

**Some projection TV system details**    For their optics, some projection TVs use three U.S. precision lens (USPL) compact delta 7 lenses. This new lens, designed by USPL, incorporates a lightpath fold or bend within the lens assembly. This is accomplished with a front surface mirror that has a lightpath bend angle of 72 degrees. Because of this lightpath bend, the outward appearance of the lens resembles, somewhat, that of the upper section of a periscope. The lens elements and the mirror are mounted in a plastic housing. Optical focusing is accomplished by rotating a focus handle with wing lock-nut provisions. Rotation of the focus handle changes the longitudinal position of the lens' B element.

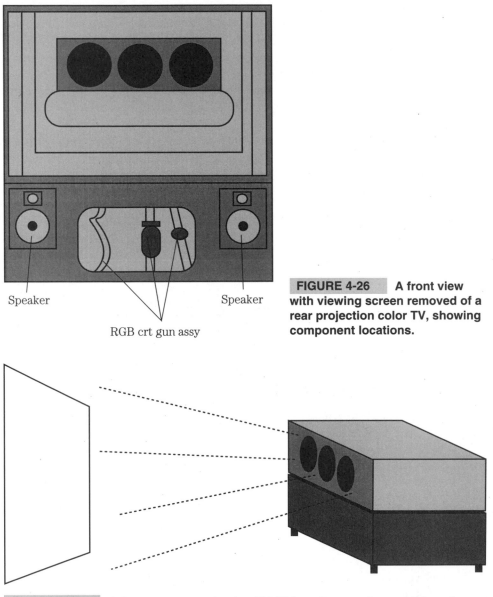

Speaker                                    Speaker

RGB crt gun assy

**FIGURE 4-26**    **A front view
with viewing screen removed of a
rear projection color TV, showing
component locations.**

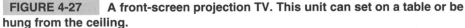

**FIGURE 4-27**    **A front-screen projection TV. This unit can set on a table or be
hung from the ceiling.**

**Projection set lightpath profile**    A side view of the TV lightpath is shown in Fig. 4-28. Note
the tight tuck of the lightpath provided by the Delta 7 compact optics. For comparison purposes, the lightpath profile of an earlier model projection set is shown in Fig. 4-29.

**Liquid-cooled projection tubes**    The rear-screen projection TVs use three projection
tubes (R, G, and B) arranged in a horizontal-in-line configuration. This type of config-

uration uses two (red and blue) slant-face tubes and one (green) straight-face tube. All tubes are fitted with a metal jacket housing with a clear glass window. The space between the clear glass window and the tubes faceplate is filled with an optical clear liquid. The liquid that is heat-linked to the outside world, prevents faceplate temperature rise and thermal gradient differentials from forming across the faceplate when under high-power drive signals. With liquid-cooled tubes, the actual safe power driving level can be essentially doubled over that of the older nonliquid-cooled tubes. This is highly desirable in terms of the large-screen picture brightness. The late-model sets use an 18-watt drive level to the picture tube, but the older-model projection sets had only an 8.5-watt drive level.

A side view of the jacket/tube assembly is shown in Fig. 4-30. The metal jacket shell extends back, well over the panel to the funnel seal and thereby functions as an effective x-ray

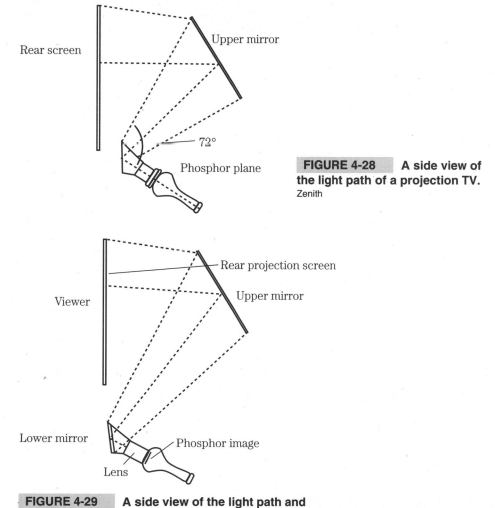

**FIGURE 4-28**    **A side view of the light path of a projection TV.** Zenith

**FIGURE 4-29**    **A side view of the light path and mirrors of a projection TV.** Zenith

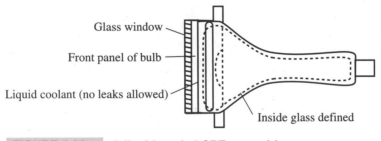

**FIGURE 4-30**    **A liquid-cooled CRT assembly.** Zenith

shield. The metal jacket also serves as the mechanical mounting and support for the picture tube assembly. The front of the metal jacket is elongated and the mounting holes are placed in the elongated sections. This is purposely done to permit the tightest possible tube-to-tube spacing for in-line tube placement.

**Optical picture tube coupling**    A pliable optical silicone separator is mounted between the glass window on the liquid-cooled jacket assembly and the rear element of the Delta 7 lens. When under mounting pressure, the silicone separator makes close contact with these two lightpath interconnecting surfaces.

**Self-convergence design**    Many large-screen projections have self-convergence and automatic convergence features. Final touch-up convergence can also be made with the remote control when in the service or set-up mode. This is accomplished in the receiver with the tilted faceplate of the red and blue tubes, in combination with shifted red and blue pointing angles, are image offsets that are used to provide for three-image convergence. This combination is required because of the shorter focal length in the Delta 7 lens design and its incompatibility with existing faceplate tilt angles. Because the receiver is a self-convergence system, registration of only the three images will be required. This is accomplished with special circuits located in the raster registration PC module.

**Picture brightness and projection screen**    Usually, the projection screen for these projection sets is a two-piece assembly. The front (viewer side) piece will be a vertical lenticular black-striped section. The rear piece is a vertical off-centered Fresnel section. The black striping not only improves initial contrast, but also enhances picture brightness and quality for greater viewer enjoyment under typical room ambient lighting conditions.

The newer-large screen receivers demonstrate increased picture brightness over previous projection TVs. This is made possible by the use of liquid-cooled projection tubes and their ability to accommodate higher-power drive signals. The improvements will be substantial and some projection sets run almost twice the brightness level as the older models. Figure 4-31 shows the location of the circuit board modules and where the projection tubes are mounted in a late-model projection TV.

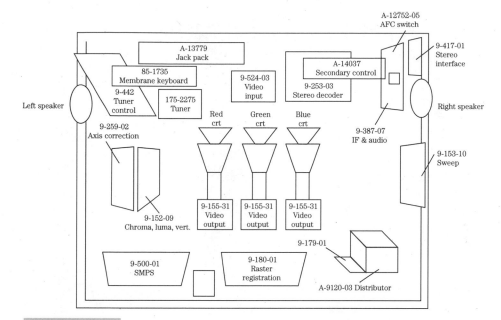

**FIGURE 4-31**    **Circuit modules and picture tube locations of a typical color TV projection set.**

# What To Do When Your TV Has Problems

Some of the TV troubles were covered in the last portion of this chapter. Some of the trouble symptoms will be photos taken from the actual TVs with the problem. Some of the other problems are within the TV.

## PROBLEMS AND WHAT ACTIONS YOU CAN TAKE

*The symptom*    The set will not operate. (No sound or picture, dark screen)
   *What to do:*

- Check the ac power outlet with an ac meter or plug in a known-working lamp. If no ac power is found, check and/or reset the circuit breaker to this outlet.
- Check the ac line cord and plug from TV to the wall outlet. Some older TVs might have an interlock plug that removes ac power from the set when the back is removed. Be sure that this interlock plug is making a good connection.
- Check and/or reset the circuit breaker on back of a TV. Other sets will have a main power fuse located on the chassis. Check fuse with a ohmmeter. Replace any blown fuse with same current (amp) rating as the blown (open) fuse. If the fuse blows again, the set probably has a shorted rectifier diode in the power supply or some other circuit is shorted or drawing too much current.
- Check the on/off switch for proper mechanical operation. Use an ohmmeter to see if the switch is working (on and off contact) electrically.

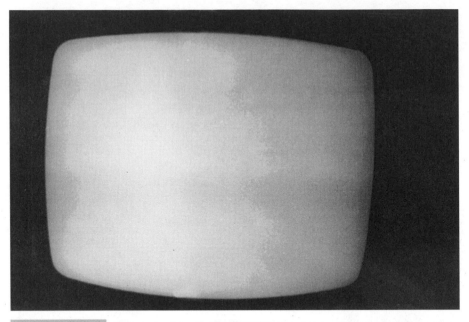

**FIGURE 4-32**    The symptom is blank (white) screen and no sound.

*The symptom*    The TV has no sound or picture. The set produces a smooth white picture (Fig. 4-32).
*What to do:*

■ Check the TV cable, antenna lead in, cable lead from DBS antenna, and be sure that all of these cable connections are good and tight. Replace the coax cable and connections if it is found defective.
■ Inside the TV is a separate tuner box that will have a shielded cable that plugs into the main chassis. Check this cable for clean, tight connections.
■ For older TVs, the tuner knob will turn and click. This indicates that it is a mechanical tuner with switch contacts. Dirty or corroded contacts can cause a loss of picture and sound. Remove the tuner cover and spray the contacts with a tuner cleaner and lubricant.

    When working inside a TV, always be very careful because high voltage is present.

■ Some TVs have a control, usually on the back, labeled *AGC (Automatic Gain Control)*. If this control is misadjusted, the picture and sound will be missing. Try readjusting the AGC.

*The symptom*    Picture width reduced (pulled in from the sides, as shown in Fig. 4-33).
*What to do:*

■ Check the dc voltage from power supply. If not correct, readjust the B+ level control if the set has one.
■ A shorted coil winding in the horizontal sweep transformer or deflection yoke could cause this problem.

*The symptom*  Very bright narrow horizontal line across the screen. This problem is caused by the loss of vertical sweep.

*What to do:*

- Check and adjust the vertical hold control.
- Check, clean, and/or adjust vertical height and linearity controls.
- Check vertical oscillator and output transistors and or IC stages.
- Check lead-wire plugs or solder connections to the deflection yoke.
- The loss of vertical sweep could also be caused by an open vertical coil winding in the deflection yoke, which is mounted on the neck of the picture tube.

*The symptom*  The picture is reduced at top and bottom (Fig. 4-34). This is also a vertical sweep problem.

*What to do:*

- Check the vertical sweep output stage components.
- It could also be a shorted winding in the vertical coils of the deflection yoke. This might show up as keystone raster shape.
- Check and adjust the vertical hold control.
- Check and adjust vertical size and linearity controls.
- Some sets have a vertical centering control. If your set has one, check and adjust it because a defective centering control will cause the picture to shrink down in size.
- Check for low dc voltages in the set's power supply and in the vertical sweep stages.

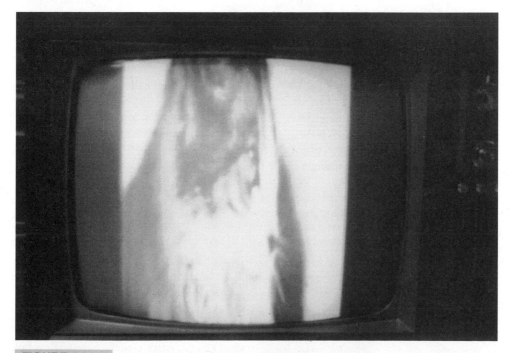

**FIGURE 4-33**    **The picture (raster) is pulled in from the sides.**

**FIGURE 4-34** **The picture pulled down from top and bottom. This is usually a vertical sweep circuit problem.**

■ Check out any of the large (electrolytic) capacitors in the vertical sweep stage or that couple this stage to the deflection yoke. To make a quick check, just bridge another good capacitor across the suspected one and see if the picture fills out.
■ A large black bar at the top or bottom of the picture tube could be caused by some type of RF noise interference. Change channels and if this black bar disappears, that is your problem. A problem in the cable system could cause this same symptom.

*The symptom* Small horizontal black lines appear across the picture and it might tend to weave (Fig. 4-35).
*What to do:*

■ The power supply might have poor low voltage regulation or faulty filter capacitors. Check B+ voltage with a meter and adjust the voltage level if your set has an B+ adjustment control. This symptom could also be caused by some type of signal interference.
■ The degaussing circuit might not be turning off after the TV warms up. Check it by unplugging the degaussing coil that goes around inside the picture tube faceplate. The thermal resistor or diode in the power supply might be defective.

*The symptom* An arcing or popping sound. This is usually around the large red HV lead and rubber cup on the picture tube. Also, in and around the HV sweep transformer stage.
*What to do:*

■ This will usually be some type of high-voltage arc. Use caution when checking out this problem. Check the large high-voltage lead (usually red in color) that goes to anode of the picture tube. Clean the rubber cup that snaps onto the CRT.
■ Check the amount of high voltage because it might be too high. You will need a special HV meter probe. Check that all ground straps around the picture tube are making good connections.

■ A blue arc in the guns (neck) of the CRT could indicate loose particles in the gun assembly or a defective tube. To clear the gun short you can carefully place the face of the tube on a flat, soft pad and gently tap the neck of the tube. This can remove any particles in the gun and clear the arc.

*The symptom*    The screen is blank except for small white horizontal lines (Fig. 4-36). The set has good sound.
   *What to do:*

■ These symptoms usually indicate a video amplifier problem. The power supply voltage and high voltage to the CRT are probably OK. Most TVs and monitors have the video board and CRT socket in one unit. This PC board will be plugged into the picture socket. Check out this video board and clean the CRT socket assembly.
■ The blank picture could also indicate a picture tube failure. A short in the CRT guns could cause this problem. The blank screen might be all one color, such as red, green, or blue.
■ In some cases, a blanking problem might cause this symptom.

*The symptom*    The picture is not clear and has poor focus (Fig. 4-37).
   *What to do*

■ Check and adjust the focus control. The control might also be defective.
■ Check the focus lead wire (large in size) and the pin on the picture tube socket.
■ Clean all pins on the picture tube socket.

**FIGURE 4-35**    Small narrow black lines appear across the screen and the picture might bend or weave.

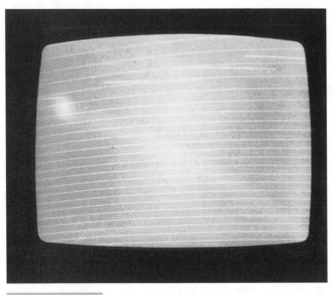

**FIGURE 4-36**    A blank (white) screen symptom, with small white lines going across. The sound is good.

**FIGURE 4-37**    An out-of-focus or blurred picture symptom.

- The focus circuit could be defective and be supplying improper focus voltage.
- The picture tube be defective.

*The symptom*    The TV has no picture or sound. Only snow and sparkles are seen on the screen. Only a hissing sound heard in the speakers.
*What to do:*

- A snowy picture is shown in Fig. 4-38. The problem could be within the TV tuner. The RF amplifier stage or input balun coils could be damaged from lightning coming into the coax cable or antenna lead wire.
- If you are using an outside antenna, the antenna or coax cable could be open or a connection could be loose or faulty.
- If you have a cable splitter and or amplifier in your home, it might have failed. These devices are used if you operate two or more TVs from the same cable or antenna.
- If you are using a DBS satellite receiver, it might not be working properly.
- If you have an older TV with a mechanical tuner, the contacts might have become dirty. You can clean them with tuner spray.

*The symptom*    Figure 4-39 shows a TV picture that rolls around and will not lock in.
*What to do:*

- Try to adjust the vertical and horizontal controls to lock the picture in. If it will not lock in, the problem is in the sync or AGC circuits. In this case, you will need a professional to repair your set.

**FIGURE 4-38**    **Picture has snow and sparkles. The sound is just a hissing noise.**

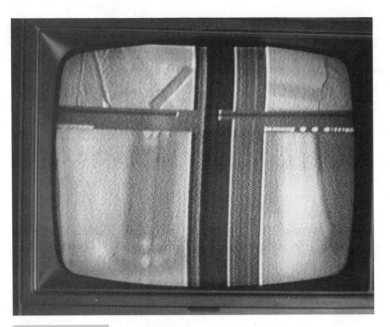

**FIGURE 4-39**    The picture is unstable, moves up and down, and tends to slip sideways. Picture cannot be locked in with horizontal or vertical hold controls. This is usually a fault in the sync and/or clipper stages.

*The symptom*    The TV has a good picture, but no sound or distorted sound.
*What to do:*

■ The speaker voice coil might be open. Check it with an ohmmeter or substitute a known-good speaker.

■ Be sure that the set's volume level is turned up and it is not in the Mute mode. This can easily be overlooked on remote TVs with screen readouts.

■ Check all wiring and plug connections that go from the TV's main chassis to the speaker. If the leads plug into the speaker, be sure that they are clean and tight. If they are soldered, the connections might have a cold solder joint. Resolder these connections, if necessary.

■ Also, check to see if any external speakers might have shorted wiring, which would cause a loss of sound or distortion.

■ For distorted audio (sound), check the speaker cone for damage or warpage, or a voice coil that might be rubbing. Replace speaker with the same impedance (ohms) as the original one.

**Conclusion**    For some of the TV symptoms and problems just covered, you will need a professional TV technician to solve or correct them. You can take a small TV into the service shop. However, the large-screen or projection TVs will need to be repaired by a professional servicer in your home.

For any TV problem, you need to find out if the TV is defective or if the signal coming into your home is either missing, substandard, or weak. A good test is to disconnect the receiver in question and connect a known-good TV. If the test set has the same symptoms, then you know that the signal into your home has a problem. You will need to call the cable company, check the outside antenna, or check the DBS system, if you are using one. If the picture is OK on the test set, you know you have a problem with your primary and/or large-screen receiver.

# VCRs: OPERATION
# AND MAINTENANCE

# Brief VCR Recording Operations

VCRs operate basically like an audio tape recorder. The video signal that your TV receives can be directly recorded by the VCR heads onto the tape by magnetic fluctuations. The signal must be processed for recording on the tape and reprocessed again when the tape is played back for the TV to reproduce a color picture. The VCR drum has two recording heads with coils of wire around them inside the drum. When one head passes over its section of tape, the other head is switched on and magnetizes that track. The head converts the processed video signal into a varying magnetic field, which occurs in the gap between the pieces of metal that the coils are wound on. Figure 5-1 shows the head drum, with the heads, coils, and the gap illustrated. Some VCRs have six heads around the drum. These machines use different sets of two heads for different tape speeds and some of the heads are also used for special video effects.

To play back the tape the VCR machine pulls the tape back around the drum again and the magnetized tape field is picked up by the heads, amplified, and processed again into a usable video signal for your TV to reproduce a color picture.

The heads on a VCR are mounted on a spinning drum where as the head of a cassette recorder is stationary and the tape is pulled past it. The spinning drum rotates at an angle to the tape (Fig. 5-2). Thus, the heads pass diagonally along the tape, laying down a slant track across the tape. This technique is called *helical-scan recording* and must be used because video signals cover a broader range of frequencies than audio signals. The speed that the tape passes around the drum will affect how good the picture quality will be. Figure 5-3 illustrates this slant track technique.

A track at the top of the tape is used for audio signal recording. The track at the bottom of the tape is the control track and it has pulses that are used to time the slant track video and keep all systems in proper synchronization as it passes around the head drum. The control track signal keeps the picture and sound in playback in good order—even if the tape stretches. A separate head is used for the control track and also to erase the tape so that it can be used again. The newer 8-mm tape does not have a separate control track, but has pilot tones that are recorded with the video signal. These pilot tones allow the 8-mm machine to optimize the video for quality picture play back.

**FIGURE 5-1**   The video head drum in a VHS recorder.

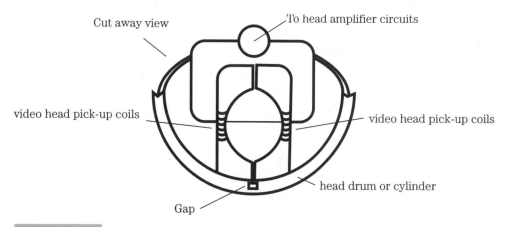

**FIGURE 5-2**   A VHS video head drum, showing the pick-up coils and gap.

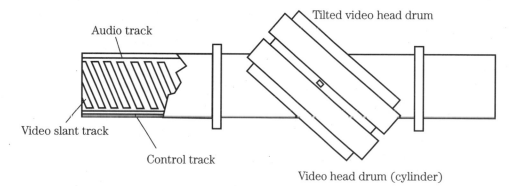

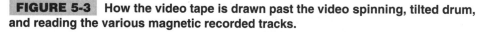

**FIGURE 5-3**   How the video tape is drawn past the video spinning, tilted drum, and reading the various magnetic recorded tracks.

As noted previously, an audio track is recorded at the top of the tape. A linear audio head picks up the audio signal for playback. The audio and control track heads operate like the heads on an audio tape recorder and pass along the tape length. In VCR recorders with hi-fi audio, the audio rides along with the video heads.

Before the video tape goes by the recording head drum, it passes by the erase head. This head is as wide as the video tape and "wipes off" any previous recording with a magnetic field. This eliminates any previously recorded sound or video pictures from mixing with the newly recorded information. Some of the newer VCR recorders and all of the Hi-8 and 8-mm recorders have a flying erase head that is mounted right on the video drum.

# Introduction to Video Recording

When the term *video tape recording* is used, it can refer to any of several systems designed to record both audio and video information on magnetic tape. To better understand how a VCR operates, first review the operations of an audio tape recorder.

## UNDERSTANDING VCR RECORDING TECHNIQUES

During the Record mode, an audio tape deck mechanically pulls the tape past a recording head. This head consists of a coil of wire and a metallic material that will conduct the magnetic fields produced by the coil. The coil receives its signal from an amplifier system, which increases the signal level enough to produce a magnetic field sufficiently strong to change the magnetic properties of the oxide materials used to coat the recording tape.

As the tape is pulled past the recording head, the magnetic properties of the oxide are changed in step with the incoming signal. To reproduce the magnetized tape information recorded on the tape, the tape is pulled past a "playback" head, which is similar in construction as the record head. In many tape decks, the same head is used for both recording and playback. The playback head is connected to the input of a sensitive preamplifier in the Playback mode. The varying magnetic field present on the tape is converted by the playback head to a changing electrical current, which closely represents the original program material. This signal is then amplified some more and fed to a speaker.

An audio tape recording system is referred to as a *longitudinal* system. The recorded material lies in a line parallel to the edges of the tape and is recorded continuously from one end of the tape to the other.

Three factors restrict the high-frequency response that a longitudinal system can reproduce:

**1** The width of the gap in the record or playback head.
**2** The size of the magnetic particles on the recording tape.
**3** The relative speed between the tape and the record or playback head.

To understand how each of these factors limit high-frequency response, consider each one separately.

Now consider the width of the head gap. Figure 5-4 shows the tape moving past the recording head gap at a fixed rate. This drawing illustrates low frequency information. Notice that the changes in the input signal are varying the magnetic flux at the head gap. The moving tape is receiving a magnetic pattern that varies with the same time relationship as

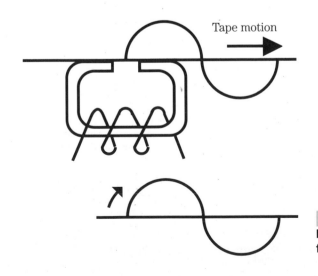

**FIGURE 5-4**  How frequencies below the crucial head-gap cutoff frequency are recorded.

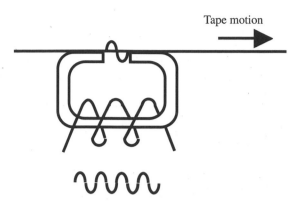

Tape motion

**FIGURE 5-5** Higher-frequency signals are cancelled before the tape moves past the gap.

the original electrical signal. But if the frequency of the signal is increased (Fig. 5-5), you will note that signals on the tape are changing so fast that the tape does not have a chance to move away from the gap before the polarity of the signal changes. The result is that the signals are canceled out before they have passed the head gap.

If the gap is made smaller, we would have a higher cutoff frequency. But, a video signal covers a range from 30 Hz (for the vertical sync pulses) through 4 MHz (for the upper sideband of the color subcarrier).

The second limitation is the size of the individual particles of oxide. Each particle of oxide can only have one level of magnetization. If the oxide particle is larger than the head gap, the constantly changing signal cannot be recorded properly. Thus, the tape oxide particles have to be even smaller than the head gap to record high-frequency information.

**Video tape recording speed**   You can increase the recorder high-frequency response by increasing the speed of the tape moving past the recording head. This spreads the high-frequency signal out over a much longer distance. The result is that a wider frequency response is possible with a relatively large recording gap. Some of the first home video recording systems used longitudinal recording formats with tape speeds as high as 100 inches per second. But these systems would "eat up" too much tape and were very impractical. A single hour of recording required 30,000 feet of tape. And, any mechanical malfunction with the tape transport would result in large piles of loose tape.

A more practical method of increasing the head/tape speed involves moving the tape at a relatively slow speed while a special spinning head is used to record the information. Early TV broadcast tape decks, for example, used four heads on a spinning disk and tape that is 2 inches wide. The heads are aligned so that one of the heads is always in contact with the tape. The head wheel is perpendicular to the tape. The tape is moving past the head wheel at a constant rate. The video information is then recorded in stripes that run perpendicular to the edges of the tape. This system is known as the *Quadraplex video recording system.*

Several factors made the broadcast Quadraplex recording system too expensive for the home video recording machine. First, the tape cost is very high and secondly, the mechanics and electronics circuitry required to control the four heads increases the cost, complexity, and size of the machines.

For these reasons, the home video system uses the low-cost helical video recording format. With this system, the video heads spin in a horizontal plane and the tape is wrapped

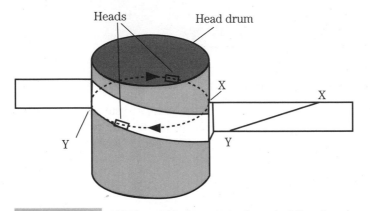

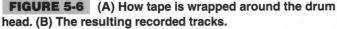

**FIGURE 5-6**  (A) How tape is wrapped around the drum head. (B) The resulting recorded tracks.

around a cylindrical drum. The tape is routed around the drum at an angle so the resulting video information is recorded in slanted lines or tracks. The spinning head produces a high tape/head recording speed with a relatively slow tape speed (Fig. 5-6).

**How the video and audio are recorded**  The video signal in helical recording systems process the luminescence (black and white) separately from the color picture information. The audio signal is not recorded in the helical VHS format, but is recorded longitudinally along one edge of the tape. A third track known as the *control track* is also recorded longitudinally. The control track provides reference pulses used to resynchronize the spinning head to the slanted video tracks during playback.

The spinning video heads are part of a mechanical PLL system. The synchronous motor that drives the head assembly is designed to operate at a speed slightly faster than required for proper playback. A fixed head picks up the control track pulses, which then "tells" the deck exactly where the slanted video tracks are recorded. Another circuit (using a magnetic sensor) determines the position of the spinning video disk head assembly, which have permanent magnets as reference points. A phase detector then determines the phase relationship of the heads to the control track pulses. The output of the phase detector controls a magnetic brake, which slows the spinning head the proper amount to keep the playback head lined up with the slanted video tracks.

# VCR Operation

The electronics of a VCR machine can be broken down to the following sections:

- The RF tuner/remote/TV channel selection section.
- The luminance or black-and-white section.
- The chroma or color section.
- The servo circuits that control the spinning record/playback heads.

- The audio processing circuits.
- The video/sound input/output handling circuits.
- The input/output circuit-handling system.

## MOVING AND LOADING THE TAPE

The tape is moved around inside the VCR with a transport system. The transport does several operations. It pulls the tape from a full reel inside the tape cassette; it then wraps the tape around the spinning drum, then past the audio/control heads, erase head, pinch rollers. After going around all of these places, the tape goes to the take-up reel in the cassette.

The first home video recorders were of the top-loading variety (Fig. 5-7). You would place the cassette into a top basket and then push it down into the VCR. This action lowered the tape into the unit's transport system. In the mid-1980s, all VCR machines went to the front tape-loading scheme.

The front-loading VCRs also have a basket that is made of plastic and metal, located behind the tape slot on the front of the unit. When the tape is pushed into the slot, a microswitch is activated. This action starts a small motor inside the VCR, pulls the cassette into the basket, and then lowers it into the transport unit. This motor and drive belt is shown in Fig. 5-8. A pin near the basket unlatches the flap on the front side of the tape cassette so that the tape can be unwound.

When the cassette is lowered, it physically contacts several parts of the transport system. Spindles move into the large, splined reel holes of the cassette. Another drive motor turns

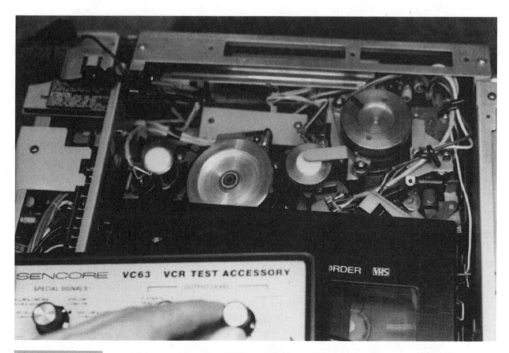

**FIGURE 5-7**   An older top-loading VCR with the case removed.

**FIGURE 5-8**   The motor and belt drive used to pull the cassette basket into the VCR to load the tape cassette.

these spindles, which move and handle the tape. Another pin goes through a hole in the cassette to release a reel brake.

On the newer 8-mm and VHS decks, a sensor lamp also goes up into the bottom of the cassette. This lamp is used to let the VCR know the tape comes to its end and stops the VCR operation. The tape blocks the light until the tape leader, which is clear, comes into range. The sensor's lamp light activates an optical sensor located at each end of the cassette. The older Sony BetaMax VCRs used a magnetic sensor and metallic strips at the tape ends to stop the machine. If these lamps burn out in the VHS machines, the unit will shut down and not operate. So, if this symptom occurs, check these bulbs.

Also, two guide posts and a capstan go behind the tape, now stretched across the cassette. These posts will pull the tape out of the cassette and load or wrap it around the cylinder drum. Then a rubber pinch roller is pressed tight to the capstan motor shaft. When the motor turns the capstan, it pulls the tape through the machine at a constant speed.

The old model BetaMax machines use the U-loader transport because it makes a U shape of the tape when it is loaded. The VHS and 8-mm VCRs used an M-loader transport, which also uses two posts to place the tape around a drum that forms the letter M shape.

The old model BetaMax VCRs keep the tape wrapped around the drum at all times and causes fast forward and rewinding to be much slower than on the VCS machines. It also causes much more tape wear.

Most modern VHS recorders use a half-loading scheme, which unwraps just a portion of the tape during Fast Forward and Rewind modes. However, the tape still touches the control head and a real-time counter can be used and the transport will work much faster.

# How the 8-mm VCR Format Works

The 8-mm video tape format was designed to meet the requirements of a smaller, compact machine size and lighter weight. The 8-mm videotape is only 8-mm wide, compared to the standard VHS video tape, which is 12.6-mm wide. As a result, the mechanism is much smaller. In the 8-mm video system, the tape runs at 14.3 mm per second and utilizes a 40-mm drum with a 221-degree tape wrap.

The 8-mm format features an FM hi-fi audio recording system. The audio signal is recorded as an FM carrier. Although most 8-mm systems can only record mono audio, provisions have been designed in the format to record stereo pulse code modulation (PCM) audio. As noted in Fig. 5-9, this gives 8-mm a far better performance than the standard VHS, with more dynamic range and a low 0.05-percent wow-and-flutter rating.

Using an advanced metal tape formulation, the 8-mm metal particle (MP) has four times the magnetic energy of the conventional cobalt ferric oxide tape used by VHS-C systems. The tape enables 8-mm to record and playback video signals with much less video dropout.

The 8-mm video format does not use a conventional control track (CCT) head, but uses the automatic track finder (AFT) circuit. With this circuit, the VCR monitors the video head position and compensates for error in the tape path. This new format and design results in a compact and low-weight high-quality recording system.

## 8-MM CASSETTE INFORMATION

The 8-mm cassette (Fig. 5-10) is a little thicker, but slightly narrower in width than a conventional audio cassette. The total tape volume is only slightly greater than an audio cassette. The 8-mm video cassette is capable of 120 minutes of recording time.

The cassette designation has also been standardized along with the cassette size. An example is the P6-120MP cassette. In the first letter-number combination, "6" represents the NTSC system using 60 fields. The next three digits, "120," indicate the record time of 120 minutes. The last two letters represent the type of tape: MP for metal powder or EP for evaporated powder.

|  | 8mm | VHS-C(SP) |
|---|---|---|
| Dynamic range | 75db | 45db |
| Frequency response line in: | 20Hz–20kHz | 50Hz–11kHz(SP) |
| Wow & flutter | 0.05% | 0.5% |

**FIGURE 5-9**  A comparison of 8-mm and VHS audio playback performance.

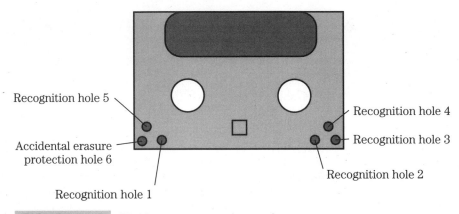

Recognition hole 5

Accidental erasure
protection hole 6

Recognition hole 1

Recognition hole 4

Recognition hole 3

Recognition hole 2

**FIGURE 5-10**   **The 8-mm cassette layout format.**

When you turn the cassette over, you will see that the tape is protected by a double door on both the outside and the inside of the tape. The door can be released by pulling back the lid lock and folding the door up to expose the tape.

A reel lock is at the bottom of the tape. When the tape is loaded into the recorder mechanism, a pin fits above the reel lock tab. When the tape is loading, the reel lock is pulled back, releasing the locking mechanism. With the reel lock pulled back, the reels are free to turn.

The 8-mm tape uses two alignment holes toward the front of the cassette to rigidly hold the cassette in position when it is seated on the loading mechanism.

The large hole in the middle of the cassette is for the light-emitting diode (LED). It works in the same way as the standard VHS recorder works. It is used to detect the transparent leader at both ends of the tape to stop the mechanism when the supply or take-up reel is empty.

Six sensor holes are in the lower-left and lower-right corners of the cassette. Three of them are now being used and more will be utilized for other formats in the future.

First is the record-proof sensor. When the record-proof switch is in the Record position, the record-proof sensor hole will appear red. If the record-proof switch is moved to the Record-Inhibit position, the red will no longer appear.

The second hole is used to sense metal powder or metal evaporated tape. When the hole is closed, metal powder tape is loaded in the cassette; when open, metal evaporated tape is in the cassette. This is used to change to equalization and other relevant recording circuits to produce optimum results with both types of tapes. The third hole is for the tape thickness.

The other three holes will be used for advanced formats and assigned to a specific function and sensors incorporated in these recorders with such new features.

## THE 8-MM TAPE FORMAT

Now check out the 8-mm tape format. Some of these techniques will apply to current recorder models and others will be used in the future.

The 8-mm video format uses the helical scan system for video recording that is used in most all video recorders built today. This system produces the high video speed with a relatively low tape speed that is required for recording the high-frequency (HF) video information. However, the 8-mm video format has been designed with many additional features, and thus the tape is divided into four areas for recording information. Two of these areas are scanned by the rotating video heads, producing the slanted tracks characteristic of all video recorders. The other two areas are conventional longitudinal tracks.

The 8-mm recorder does not have fixed heads for recording or playback of longitudinal information. All information recorded onto the tape is accomplished with the rotating video head. An 8-mm tape format layout is illustrated in Fig. 5-11.

The first and largest area contains the video, both luminance and chroma, the FM mono audio signal, and the tracking signal. In the 8-mm format, the major portion of the information is recorded in this area. The remaining three areas are optional and are used for special features.

In this format, two video heads are used, and they are distinguished from each other by a ±10-degree azimuth difference to reduce crosstalk between the adjacent channels. The width of the tape track is 20.5 microns.

The second area of information extends for 1.25 mm below the video track. This area is for the pulse code modulation (PCM) audio and tracking signals, and is reserved for the recording of stereo PCM-encoded audio. It is not a requirement of the 8-mm format that all units play PCM audio, but this portion of the tape must be available on all 8-mm recordings. A tape recorded with PCM audio information will also have the FM audio mixed with the video signal. This tape can be played on any 8-mm machine without the PCM capability, which will produce sound from the FM audio. In addition to the PCM audio signal, the tracking signal is also recorded in the same area of the tape.

**Rotational head drum operation**  The recording of these two areas of information by the rotating video heads is made possible by scanning the tape for approximately 221

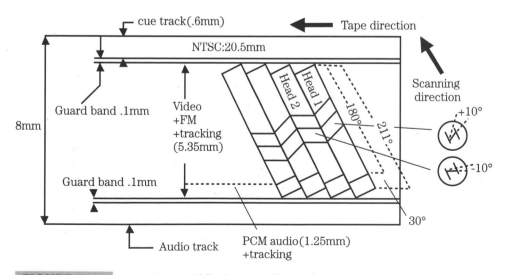

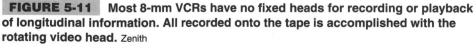

**FIGURE 5-11**  Most 8-mm VCRs have no fixed heads for recording or playback of longitudinal information. All recorded onto the tape is accomplished with the rotating video head. Zenith

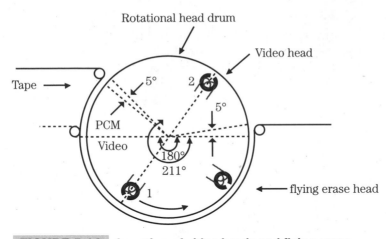

**FIGURE 5-12** **Location of video heads and flying erase head on an 8-mm machine.** Zenith

degrees (Fig. 5-12). In some 8-mm machines, the heads scan only 180 degrees of the tape because that's all that is necessary for one video head to be in contact with the tape at all times. For 8-mm machines with PCM audio, the tape is wrapped an additional 40 degrees. During this additional tape contact, the entire audio signal is recorded as PCM information, in addition to video information.

The third area of information is an optical cue track, 0.6-mm wide at the top of the tape. This cue track is separated from the video tracks by a 0.1-mm guard band. This cue track is optional, like the PCM audio area. It can be used for future format innovations to record editing information. Like the PCM signal, this cannot be recorded without also recording the full video, FM audio, and tracking signals in the scanning portion of the tape. So, any tape made with or without a cue signal can be played back on any 8-mm video tape machine.

The fourth area of information on the tape is a longitudinal audio track at the bottom of the tape. This track is 0.6-mm wide and, like the cue track, is separated by a 0.1-mm guard band. Also, like the cue track, this longitudinal audio track is optional. A camcorder that uses this track must still record using the FM audio mixed in with the video and tracking signal on the same tape. This is done so that all tape units will be able to play 8-mm tapes no matter which machine it was recorded on.

The RF switching pulse is used to switch between the video heads and produce a continuous RF envelope. In the 8-mm format, with the additional wrap for PCM audio, the RF switching pulse is still used to select video information from the heads. However, this switching pulse occurs between the PCM and video information, in the same relative position as it would occur in a conventional video recorder.

The PCM audio information is pulled off the tape by a different select pulse.

**How the 8-mm tape is erased** Some 8-mm machines do not use a full erase head to remove previously recorded information from the tape. Some units use a flying erase head, such as those found on professional editing machines. The flying erase head is positioned on the video drum assembly. For more 8-mm recorder operations, refer to Chapter 7, Camcorder Operations.

# Some Video Recording Basics

Let's now look at some video recording basic techniques. Just like audio tape recording, video information is also stored on magnetic tape by means of a small electromagnet or recording head. The two poles of the head are brought very close together, but they do not touch. This creates a magnetic flux that extends across the separation of the gap.

If an ac signal is applied to the coil of the head, the field of flux expands and collapses according to the rise and fall of the ac signal. When the ac signal reverses polarity, the field of flux will be oriented in the opposite direction and continue to expand and collapse. This changing field of flux is what accomplishes the magnetic recording on the tape. If this flux is close to a magnetic material, it will become magnetized according to the intensity and orientation of the field of flux. The magnetic material used is oxide-coated (magnetic) tape.

As an example, the tape has differently magnetized regions that can be called north (N) and south (S), in proportion to the ac signal. When the polarity of the ac signal changes, so does the direction of the magnetization of the tape, as shown by one cycle on the ac signal (Fig. 5-13). If the recorded tape is then moved past a head whose coil is connected to an amplifier, the regions of magnetization on the tape will set up flux across the head gap that, in turn, induces a voltage in the coil to be amplified. The output of the amplifier is then the same as the original ac signal. This is basically what is done in audio recording, with other methods for improvement, such as bias and equalization.

Some inherent limitations in the tape recording process affect videotape recording. Figure 5-13 shows that the tape has north and south magnetic fields that change according to the polarity of the ac signal.

If the speed of the tape past the head (head-to-tape speed) is kept the same, the changing polarity of the high-frequency ac signal would not be faithfully recorded on the tape (Fig. 5-14).

As the high-frequency ac signal starts to go positive, the tape starts to be magnetized in one direction. But the ac signal quickly changes its polarity, which will be recorded on most of the same portion of the tape. North is covered by south magnetic regions, which

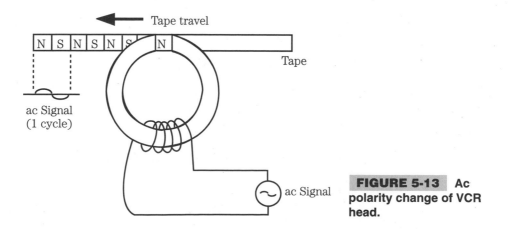

**FIGURE 5-13**  Ac polarity change of VCR head.

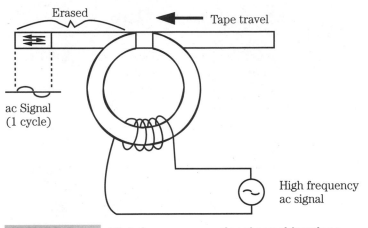

Erased

Tape travel

ac Signal
(1 cycle)

High frequency
ac signal

**FIGURE 5-14** **High-frequency ac signal considerations.**

results in zero signal on the tape (this means that it is self-erasing). To keep the north and south regions separated, the head-to-tape speed must be increased.

To accomplish this for video recording, the video heads as well as the tape must be moved. If the heads are made to move fast across the tape, the linear tape speed can be kept very low. In two-head helical recording, the video heads are mounted in a rotating drum or cylinder, and the tape is wrapped around the cylinder. This way, the heads can scan the tape as it moves along. When a head scans the tape, it is said to have made a track. This two-head cylinder and tape tracking is illustrated in Fig. 5-15.

In two-head helical format, each head records one TV field (262.5 horizontal lines), as it scans across the tape. Therefore, each head must scan the tape 30 times per second to yield a field rate of 60 fields per second.

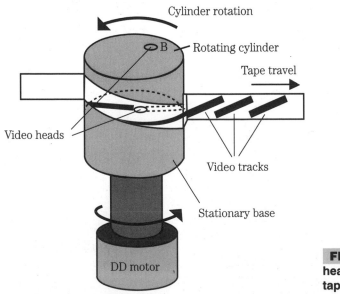

Cylinder rotation

B — Rotating cylinder

Tape travel

Video heads

Video tracks

Stationary base

DD motor

**FIGURE 5-15** **Video head scan and recorded tape tracks.** Zenith

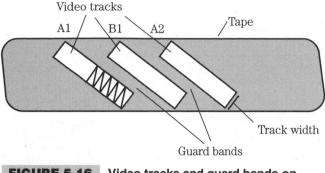

**FIGURE 5-16**   **Video tracks and guard bands on the tape.**

Please note in Fig. 5-15 that the tape is shown as a screen wrapped around the head cylinder to make it easy to see the video head. A second video head is 180 degrees from the head shown in front. Because the tape wraps around the cylinder in the shape of a helix (helical), the video tracks are made as a series of slanted lines. Of course, the tracks are invisible, but it is easier to visualize them as lines. The two heads "A" and "B" make alternate scans of the tape.

An enlarged view of the video tracks on the tape is shown in Fig. 5-16. The video tracks are the areas of the tape where video recording actually occurs.

Another aspect of video recording is that magnetic heads have characteristics of increased output level as the frequency increases, which is determined by the head gap width. In operation, the lower-frequency output of the heads is boosted in level to equal the level of the higher frequencies. This process is also used in audio recording and is referred to as *equalization*.

## VHS RECORDER SEARCH MODES

Some other VHS functions include search-forward (cue) and search-reverse (review). To quickly find a particular segment on a recorded tape during playback, you can speed up the capstan and reel tables to nine or more times the normal playback speed, either forward or reverse, by pressing the cue or review buttons. At this time, noise bars will appear across the screen because of head crossover. This is normal on some VCRs. For example, some models show four noise bars in Cue mode and five noise bars in the Review mode. The bars for the Cue and Review modes are shown in Figs. 5-17 and 5-18.

## SOME VIDEO HEAD CONSIDERATIONS

A reduction in track width requires the use of smaller video heads. But just making the heads smaller does not actually make them better. With less actual head material to work with, the magnetic properties of the head will suffer. To compensate, a change in head material is necessary. Because the VHS recorders and camcorders need to be small and light-weight, a reduction in the size of the head cylinder is needed.

A reduction in the size (diameter) of the head cylinder changes the head-to-tape speed. Keep in mind, that the head-tape speed affects the high-frequency recording capability of the head. To offset this problem, the head gap size will have to be reduced.

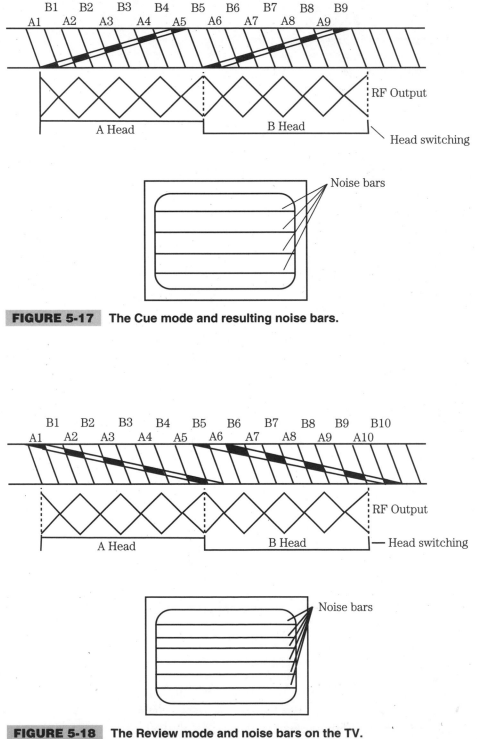

**FIGURE 5-17**  **The Cue mode and resulting noise bars.**

**FIGURE 5-18**  **The Review mode and noise bars on the TV.**

The use of hot pressed ferrite as video head material in the VHS recorder helps improve the characteristics of the smaller size heads. The hot pressed ferrite also has uniform domain orientation that further improves the head characteristics. It has been proven in many actual operating test runs that the hot pressed ferrite material produces a superior video head.

From the preceding explanation, the need for smaller head gap size has become very apparent. In VHS recorders, the video head gap width is a mere 0.3 micrometer. This is quite a contrast from ordinary video heads used in other helical applications whose head gap widths are typically in the micrometer size.

## AZIMUTH RECORDING

*Azimuth recording*, a term used to define the left-to-right tilt of the head, is used in the VHS format to eliminate the interference or crosstalk picked up by a video head. Figure 5-19 illustrates this tilt. Again, because adjacent video tracks touch, which induces cross talk, a video head can pick up some information from the adjacent track when scanning. The azimuth of the head gaps ensure that video head A only produces an output when scanning across a track made by head A. Head B, therefore, only produces an output when being scanned by Head B. Because of the azimuth effect, a particular video head will not pick up any crosstalk from an adjacent track.

## HOW TO CLEAN AND CHECK OUT YOUR VCR

In many ways, the VCR is not much more complicated for routine cleaning than some cassette decks. However, some of the electronic and mechanical tape transports are considerably

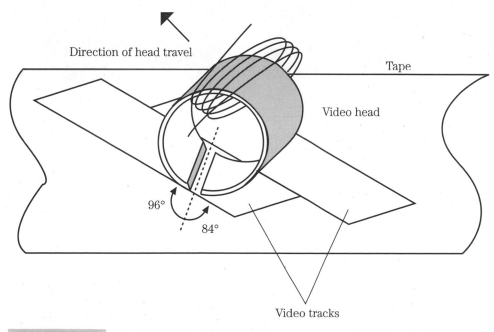

**FIGURE 5-19**   The VHS video head format.

more sophisticated. For this reason, cleaning the cylinder head, tape guides, and belt drives are more involved than for an audio cassette machine.

In an audio recorder, the magnetic tape passes over stationary heads to record and play back sound. The VCR, by comparison, has rotating video heads along with fixed audio, control track and erase heads. The rotating heads are needed because of the higher frequencies. The tape must travel across the heads at a much faster rate than for audio recording.

**VCR cleaning tapes**   As with any sophisticated machine, the VCR requires proper care, and periodic maintenance and cleaning to keep it operating at peak effectiveness.

Some service centers have reported that up to 90% of the VCRs returned to them for repairs only needed a thorough cleaning of the tape guides and video heads. The most important concern is the proper maintenance and cleaning of the delicate record and playback cylinder heads.

For this reason, a number of different types of head-cleaning systems are available at various electronic parts stores and service centers for you to use. Some use liquid solvents and others use a dry abrasion technique.

There are two ways for you to clean your VCR. It can be taken apart and cleaned manually with solvents, etc., or you can use a cleaning tape. There are wet and dry types of cleaning tapes. The dry tape cleaners (Fig. 5-20) do not clean as well as manual cleaning and can wear the head out quicker because they are abrasive. The wet-tape cleaners use a cleaning fluid and can cause other problems in your VCR.

**FIGURE 5-20**   **A BetaMax dry head-cleaning tape.**

A dry tape cleaner is inserted into the VCR; watch the TV screen or listen for an audio tone or voice cue. Some will tell you when to stop and others will actually eject the tape when the cleaning is completed. For a very bad "head clog" problem, you might have to run the cleaning tape two or three times. If this will not clean the video head, you will have to clean them manually.

**Cleaning the VCR mechanism** Caution: Before removing the top cover of your VCR to clean the heads, make any adjustments or perform minor repairs be certain that you have unplugged the VCR from the ac outlet.

To remove the top of the VCR or the case, use a small screwdriver to take out the screws. Now put the screws in a small plastic bag or cup so that they will not be lost. Some of the screws might be of different sizes, so you should note where they were removed. Also, do not drop any screws or other tools into the VCR mechanism.

Some VCRs will have a circuit board or metal cover over the head cylinder and transport. This must be removed before you can clean the heads, capstan, and tape guide tracks. You are now ready to clean the heads, tape guides, capstan, and pressure and pinch rollers.

**VCR cleaning tips and materials** You will need chamois head-cleaning sticks. Only use cotton swabs for cleaning other parts of the machine, special head-cleaning fluid or alcohol, and a clean, lint-free cloth.

Even with normal operation, contaminants in the air can cause a VCR to perform below par. Particles of dust, smoke, loose magnetic tape oxide, and oxide binder could, over a period of time, cause a hard buildup on the head surfaces.

This buildup causes the tape to be physically spaced away from the proper firm contact with the face of the cylinder head, which reduces the amount of signal being recorded and played back. This signal loss, because of head buildup, will cause distorted low audio with a noticeable lack of high frequencies. There will be a loss of overall picture clarity and also snowy picture with lines. It will also be necessary to clean tape oxide buildup from all the contact points in the tape path, such as rollers and guides.

In extreme cases, where tape oxides are allowed to build up over a long period, the accumulation can become large and ragged enough to scratch or tear tapes or interfere with the precise tape speed required for VCR operation. Regular VCR cleaning is the only way to prevent oxide buildup and other major machine problems.

The rotating video heads are the most delicate and expensive parts of your VCR. They must be treated with the utmost care and kept very clean. The video heads actually penetrate into the tape oxide when recording and being played back; thus, a small amount of oxide will always be shedding from the tape onto the heads and guides in the normal VCR operating process.

Oxide buildup on guides and heads can cause very poor recording and playback of your video tapes. This will appear as streaks or noise (Fig. 5-21) or if built up enough, no picture at all. In fact, dirty heads will show up first in the Play mode because of the relatively weak magnetic field from the tape that must be transferred to the head drum. Another snowy picture and streak condition you might think is caused by dirty heads is actually tape dropout. This occurs when a few horizontal lines across the picture are missing (Fig. 5-22) and appear as a line of interference. This dropout is caused by a streak of oxide missing from the tape or a very worn tape.

**FIGURE 5-21** Streaks across the picture, caused by oxide buildup on the video heads and guides.

**FIGURE 5-22** Poor picture quality and lines across the TV, caused by poor tape drop-out conditions.

**Cleaning the VCR**  To clean the video heads, use a special cleaning pad, such as chamois cloth or cellular foam swabs. If these are not available, lintless cloth or muslin can be used. Cotton-tipped swabs should not be used for cleaning the cylinder heads. The cotton strands can catch on the edges of the video heads and pull the small ferrite chip away from its mounting and ruin the head. However, cotton swabs soaked in cleaning fluid can be used to clean tape guides, control track, and audio and erase heads. To clean the heads and guides, use methanol or isopropyl alcohol.

The cleaning pad should be liberally soaked in the cleaning fluid (alcohol), then gently and firmly rubbed sideways across the heads. Never rub it up and down because this action might damage the heads. Clean the whole head in this same sideways motion. Be sure that you then clean all places that the tape touches. Do not touch these parts with your fingers because the oil from your skin will attract dust and dirt. However, it's a good idea to hold the cylinder on top so that it will not rotate as you clean (in the direction of head rotation). This correct cleaning technique is illustrated in Fig. 5-23. The most important aspect of head cleaning is to be very careful. The more often you clean the heads, the easier it will become.

If you find a head that is very dirty and the head chip is plugged up with oxide, soak the area around the head chip with alcohol two or three times. Then use an old toothbrush, also soaked in alcohol, to clean out around the head chip. Do this very carefully because the head can be damaged. You might want to cut the bristles of the toothbrush shorter. A spray head cleaner can also be used to clean VCR parts, etc.

**VCR cleaning steps**  Here are the steps for cleaning a VHS VCR. Be sure that the power is disconnected from the VCR before removing the case and cleaning.

1  First, remove the screws that hold the top cover on so that it can be removed. These are usually Phillips screws. Be careful not to damage the screw heads.
2  The two (some VCRs have four or more heads) very delicate video heads are located on the rotating head cylinder. Gently rotate this cylinder to bring each head into a position for cleaning. You might want to use an inspection mirror to get a better look at the head faces. Do not actually touch the highly polished face of the disc cylinder with your fingers. The control track, audio head, and erase head are located on each side of the head cylinder. A portion of the cylinder head, control track and audio heads are shown in Fig. 5-24.

To clean these heads, first saturate one of the cellular foam cleaning swabs with a good tape-head cleaner fluid. Clean the heads using only a horizontal (side-to-side) motion.

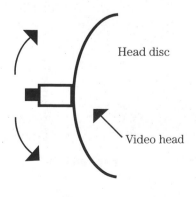

Head disc

Video head

**FIGURE 5-23**  The correct rubbing technique for cleaning the video heads.

**FIGURE 5-24**    The cylinder head, control track, and audio head locations in an VHS recorder.

**FIGURE 5-25**    The proper technique for using a swab to clean a VCR cylinder head.

To ensure that cleaning is done in a side-to-side fashion, hold the swab stationary against the head and use the cylinder head to rotate it back and forth.

Do not clean with a vertical (up-and-down) motion. This could damage the fragile heads. The photo in (Fig. 5-25) illustrates how the cylinder head should be cleaned.

**3** The control track and audio heads should be cleaned in the same way that you have cleaned the video heads (using only a horizontal scrubbing motion).
**4** After all the heads have been cleaned, perform the same cleaning functions on all contact places (rollers, guides, etc.) that the tape travels around.
**5** Now replace the top cover of the VCR and install all of the screws. Wipe off the VCR case and the covers with a clean anti-static cloth.

Do not use any lint-producing cloth or materials because the lint might be left in the machine during the cleaning process and could cause wear and other damage to the delicate VCR components.

An expanded view of the cylinder head assembly is shown in Fig. 5-26.

Be sure that the tape heads and tape path in the VCR are completely dry after the cleaning procedure before operating it. Otherwise, the tape might not thread properly and jam up some portion of the transport system. You might want to use Nortronics VCR-103 spray tape head cleaner because the cleaner evaporates completely, leaving no oil or other residue.

**Cleaning the belts and drive wheels**   All of the drive belts, pinch rollers, and drive wheels should be checked and cleaned when you have the VCR apart for head cleaning. Check for loose drive belts and worn rubber drive wheels. Some of these gears and rubber drives are shown in Fig. 5-27. If only the rewind and fast forward operate properly, then suspect that the drive wheels are worn. The drive belts and wheels can be cleaned with isopropyl alcohol or any type cleaner made for this purpose. If new belts are installed, be sure that they are put on properly and check them for proper tension. Always make a careful check when drive belts, drive wheels, gears, or other mechanical parts are changed or adjusted. One such gear assembly is shown in Fig. 5-28.

Much of the VCR's mechanical alignment and parts replacement require using special jigs, gauges, and fixtures for correct tape tension and clutch/brake operations. These repairs must be made by a trained VCR service technician.

**Video head "stiction"**   VCRs that have logged hours of use might begin to exhibit a condition described as "stiction." The word, a combination of the words "sticking and friction," indicates a condition of the video head drum assembly that causes the tape to stop moving during the Record or Play modes. If this occurs and continues unchecked (as in recording with the programmed time mode), severe clogging of the video heads and tape damage could result.

The apparent cause of stiction is the loss of an air cushion between the tape and the record drum head cylinder. As the VCR is used, friction from tape travel polishes the drum

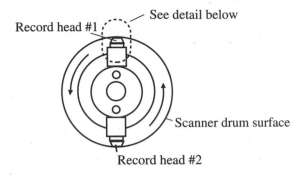

Top view of rotating record-head assembly

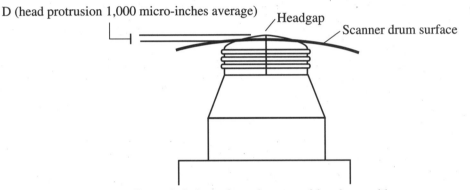

Expanded view of rotating record-head assembly

**FIGURE 5-26** The video heads are mounted in a rotating drum assembly. The heads protrude slightly from the drum, which allows them to push into the tape coating. This design provides close contact between the head and tape, which is needed to record the high-frequency video information.

surface smooth. This prevents the required air buildup, and the tape adheres to the drum. The only sure cure for this problem is to replace the head drum assembly.

**Head cleaning review and tips** Before you take apart the VCR to clean the heads, try running a cleaning tape. If you cannot clean the head this way, then use a cleaning patch. Coat the cleaning patch with alcohol or head-cleaning fluid to the area indicated in Fig. 5-29. Touch the cleaning patch to the head tip and gently turn the head (rotating the cylinder) right and left. Do not move the cleaning patch vertically and be sure that only the buckskin on the cleaning patch contacts the head. Otherwise, the head might be damaged.

Next, thoroughly dry the head. Then run an expendable tape. If alcohol or head-cleaning fluid remains on the video head, tape guides or rollers, the tape might be damaged when it contacts the head surface.

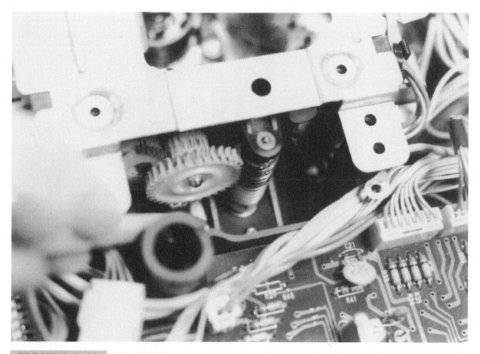

**FIGURE 5-27**    The VCR gears and gear driver that might need to be cleaned and lubricated for smooth operation.

**FIGURE 5-28**    Always be sure that the tape hub drive gears are clean and turn smoothly.

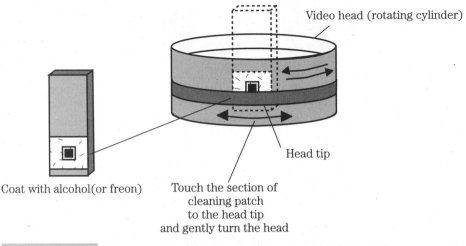

Video head (rotating cylinder)

Head tip

Coat with alcohol(or freon)

Touch the section of
cleaning patch
to the head tip
and gently turn the head

**FIGURE 5-29**   **The proper technique for using the cleaning patch.**

Now, clean the tape-transport system, drive system, etc., by wiping it with a cleaning patch wetted with alcohol or cleaning fluid. Notice the tape transport system contacts the running tape. The drive system consists of these parts that move the tape.

# VCR Connections

Before you connect the VCR and TV, check the following:

**1** Turn off the TV and/or unplug it from the ac outlet.
**2** Turn all connectors tight and firm. A loose connection could cause a distorted picture.
**3** Connect the ac power cords after all connections of the video cassette recorder and TV have been completed.
**4** After connecting the cables, set the switch on the VCR to channel 3 or 4 (whichever channel is not used in your area for a local TV station).

Now look at the various ways that you can connect your VCR and TV.

Figure 5-30 illustrates how the outside antenna is connected to your VCR and how the coaxial cable is then connected to the TV. This is a very basic VCR/TV hookup.

Figure 5-31 shows a direct cable hook-up to your VCR and TV without a converter/descrambler box. With these connections, you can:

**1** Use your VCR remote control to select channels.
**2** Program one or more unscrambled channels for unattended recording.
**3** View one channel while recording another channel.

With the VCR connections shown in Fig. 5-32, you can record and view any channel, including scrambled channels. However, channel must be made with the cable company converter box. This means you cannot change channels with your VCR's remote control.

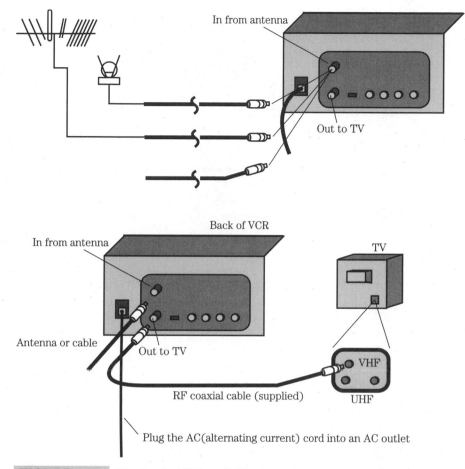

In from antenna

Out to TV

Back of VCR

In from antenna

Antenna or cable

Out to TV

TV

VHF

UHF

RF coaxial cable (supplied)

Plug the AC(alternating current) cord into an AC outlet

**FIGURE 5-30**    **The basic VCR and TV connections when hooking-up an outside antenna.**

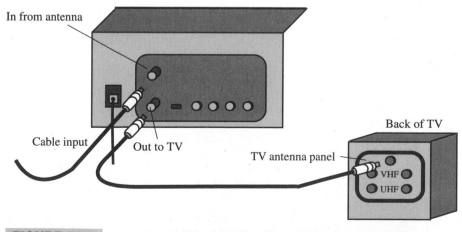

In from antenna

Cable input

Out to TV

Back of TV

TV antenna panel

VHF

UHF

**FIGURE 5-31**    **How the cable is connected to a VCR when a cable converter/descrambler box is not used.**

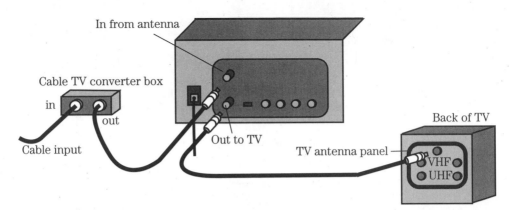

**FIGURE 5-32** The cable connection to a VCR and TV when using the cable-TV converter/descrambler box.

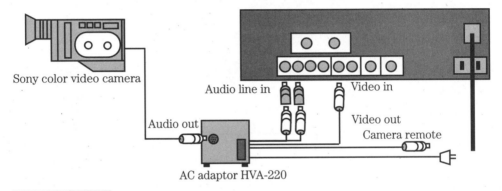

**FIGURE 5-33** The cable connections and hook-ups for using a video camera to record on your VCR.

Unattended recording is limited to one channel at a time, and you cannot view another channel other than the one you are recording.

Figure 5-33 illustrates the cable connections needed for connecting your video camera to the VCR to make your own videos.

## S-VIDEO CONNECTIONS

Some VCRs and TVs have S-Video input and output connections. These take a special-type connector and cable. This system requires a special S-Video cable and another cable for the audio signal. The audio jacks are of the RCA type. The S-Video connections result in much better picture quality because the signal does not pass through the RF tuner and IF circuits of the TV.

S-Video outputs are used in Super VHS, Hi-8, and BetaMax recorders. Many TVs and DBS receivers have S-Video features. The S-Video outputs carry the brightness and color parts of the video signal with no audio, but produce a much clearer and sharper picture. If your TV has an S-Video input, connect the S-Video cable from the VCR's S-Video output jack to the TV's S-Video input jack. You must also connect the audio cable and/or two cables for left and right stereo sound. If your TV has no S-Video input, just ignore the S-Video output jack on the VCR and use the standard video output connections.

# Common VCR Problems and Solutions

This section covers some of the common VCR problems. Most VCRs work for many hours each week. If you record an hour-long TV program and then play it back, you have actually operated the VCR for two hours. And when you rent a movie, about 1000 feet of tape is run through the tape transport mechanism. When you rewind the tape, more hours of wear and tear are added to the machine. For all of these operations, the rubber belts, pinch roller, idler wheels, and head drum are working when the tape is moving and wearing out, which causes many of these common VCR problems. In most cases, the mechanical VCR problems are caused by rubber parts wearing out or hardening with age.

## COMMON VCR PROBLEMS

Here are some of the common VCR symptoms that indicate that various rubber parts have become worn.

- Tearing up the tape.
- VCR will not rewind.
- Tape seems to slip.
- Tape only loads part way.
- VCR will not fast forward.
- Intermittent machine shutdown.
- Portions of the tape are damaged.
- Take-up reel will not turn.
- Squeaking sounds occur when tape is running.
- VCR will not operate in any mode.

Now look at some of these VCR symptoms and see what some of the probable causes could be.

**Tearing or "eating up" the tape**   If the tape is still hanging out when the cassette is ejected, the tape will be torn or damaged when the flap closes on the tape as it comes out of the machine. This usually is caused by a worn-out or damaged idler tire. Check and replace any broken parts.

**Tape only loads part way**   If this problem occurs, the VCR will start, then shut down. You might see that the cassette will go into the machine, hear it drop into place, and then shut down. Take the top cover off of the machine and load another cassette. You might see the guide posts trying to load the tape around the drum, and then stops half way around. The loading motor might keep running.

The problem is usually caused by a loose or worn loading drive belt. Check and replace the belt. Also, the guide post tracks should be cleaned and lubricated at this time.

**Tape will not rewind**  This problem could also be a slow rewind condition. This will usually be caused by a worn idler tire. On some VCRs, this condition might cause the machine to shut down.

**Tape slippage**  When tape slippage occurs, the symptoms will be an intermittent change of tape speed. This will be noticed in the sound first and then the video will be affected. You will hear a pitch change in the audio.

This slippage is almost always caused by a worn or very dirty pinch roller. The pinch roller and the capstan is what pulls the tape through the VCR at a constant speed.

**Squealing noises**  Squealing noise is usually caused by worn drive or loading belts. Also, a clutch could cause this sound. When checking belts, look for a shiny appearance because this indicates that it needs to be replaced. At times, the squealing might be caused by a dry capstan shaft or motor bearing that needs lubrication.

**Take-up reel turns slowly or not at all**  On most VCRs, when the take-up reel stops, it shuts the machine down. With the VCR top removed, see if the tape is bunching up after the pinch roller and capstan. This problem can usually be solved by replacing the rubber idler wheel. Some of the newer VCRs use a gear drive for the idler. With a worn idler tire, the take up reel cannot turn, so the sensor shuts the VCR down. Many intermittent VCR shutdown problems can be traced to dirty or worn belts, idler tires, and pinch rollers.

**Pinch roller problem**  If the edge of the tape is damaged (looks like it is curled), the most likely problem is a worn pinch roller.

**VCR troubles reviewed**  When you have a VCR problem, take a quick review of the following troubles and solutions:

- *Slipping capstan drive belt*  Some older VCRs use a rubber belt to connect the capstan motor to the capstan drive shaft. Check to see if the belt has stretched from age. Also be sure that the belt is clean and dry. Oil from other mechanical parts of the VCR can get on the belt and cause slippage as the oily part of the belt contacts the smaller capstan motor pulley. Also, check and see if any foreign material has become lodged on the belt or the pulley.
- *Worn pinch rollers*  All VCRs use a rubber pinch roller to put pressure between the tape and capstan. The rubber idler can become hardened and glazed with normal usage. This wear puts inconsistent pressure on the tape and capstan, which causes slippage. The pinch roller can also become coated with oxide from the tape. This will often cause the tape to momentarily stick to the pinch roller and produce erratic tape movement.
- *Capstan oxide buildup*  Good contact between the capstan and the tape is crucial for good operation of your VCR. Capstans often become coated with loose oxide from the tape. This reduces the gripping ability of the capstan and causes the tape to slip. Be careful to always remove any oxide coating on the capstan shaft.

**FIGURE 5-34**   The capstan drive motor in a VCR. Note clip-in fuses above the motor. If the VCR is dead, check for a blow fuse. If the tape gets tangled or the cassette jams, the fuse might blow because of heavy current being drawn by the motor.

■ *Oxide buildup on the tape path*   Oxide material from the tape can also build up anywhere along the path the tape travels. This buildup produces tape drag. If the buildup becomes excessive, it increases the tape drag until the tape slips on the capstan.

■ *Tape tension problems*   Excessive back tension on the tape produces the same results as an oxide buildup. It creates a drag on the tape that ultimately causes the tape to slip on the capstan. A tape-tension gauge will have to used for this adjustment.

■ *Take-up reel drive clutch problems*   After many hours of use, the take-up reel's felt clutch assembly becomes worn, causing it to grab or "chatter," instead of slipping smoothly. This causes the tape speed to fluctuate at such a high rate, the capstan servo circuits cannot correct the problem. The clutch will need to be replaced.

**Capstan motor problem**   A bad bearing, defective capstan motor winding, or faulty drive signal can also cause erratic tape movement. The capstan motor is shown in Fig. 5-34. To check this motor, you need to manually turn it. Feel if it is catching as you rotate the capstan motor. A sudden catch is an indication of a bad motor bearing. Replace the motor because most have sealed bearings and cannot be lubricated.

**FIGURE 5-35**    **The Trackmate video tape recorder cleaning cassette.** Trackmate

# Trackmate Cleaning System

The Trackmate Video Hyper Brush is a maintenance cassette using three, absorbent fiber brushes, which are 175-percent wider than a video tape. The 60,000 flexible, absorbent filaments seek and remove dirt, dust, and oxide particles, without causing the abrasion common to fabric tape cleaners. The Trackmate cassette is shown in Fig. 5-35. To use this cleaner, you apply pure isopropyl alcohol, from the applicator pen, to the brushes. Putting the cassette into the machine, you then press play and the Hyper Brush fully extends into the unit, form fitting to the surface of the delicate head/drum assembly. Within 30 seconds, the damaging dirt is safely removed from the gaps and grooves. To avoid the recontamination common to other cleaning products, the Video Trackmate cleaner automatically counts the number of cleanings and alerts you when to remove and clean the brushes.

The Trackmate system includes the VHS video head cassette, the VHS-C camcorder care kit and the 8-mm camcorder care kit.

For more information and dealers, contact:

Trackmate
5209-B Davis Blvd.
North Richland Hills, TX 76180
(800) 486-5707

# 6

## DIRECT BROADCAST

## SATELLITE (DBS)

## SYSTEM OPERATION

# Introduction to Satellite TV

At this time, several direct digital satellite TV systems are in operation around the world. This chapter shows how these systems work and gives you information on various items you can check if the receiver and dish do not work. You might also want to obtain another of my books from "McGraw-Hill" that has complete instructions for installing one of these 18 inch direct TV dishes and various troubleshooting information.

**FIGURE 6-1    The satellite dish mounted on an antenna mast.**

These TV satellites or "birds," as they are often called, revolve around the earth at over 22,000 miles in a geosynchronous orbit which makes it appear that they are not moving. These TV satellites pick up signals with their receivers and then send the video signals via onboard high-power 120-watt transmitters back down to earth in a pattern that covers all of the 48 main land states. The signal is strong enough to be picked up with a small 18-inch dish that is shown in Fig. 6-1. These TV satellites operate like an amateur radio repeater.

In the geosynchronous orbit, the satellite is placed over the equator at approximately 22,300 miles above the earth. A satellite in this type of orbit will not wander north or south and will have an earth-day rotation. This satellite in the sky will appear to stand still in a fixed position because its speed and direction matches that of the earth's rotation.

The uplink transmitter station pointing at the satellite in a geostationary orbit, and the downlink to your dish will not require tracking equipment because the earth's rotation matches that of the satellite.

## KEEPING THE SATELLITE ON TRACK

Because the earth's gravitational pull is not the same at all places as the satellite rotates around it and the moon also affects its position in space, the satellite is always being pulled off course and must be corrected.

Position and attitude controls are used counter these gravity pulls and keep the satellite in its proper slot. These adjustments are accomplished by on-board rocket thrusters that are fired to obtain course corrections. In fact the lifespan of the satellite is determined by how fast these thrusters use up the fuel for stabilization. Once the fuel is used up, the satellite will wander off course and become unusable.

In the early days of satellites, the spacing between them was four degrees. Now, with much improved antenna directivity, the satellites can be placed at 2-degree spacing.

## POWERING THE SATELLITES

Because the satellite is not a passive device, it has to have the ability to collect and store electrical energy.

Solar cells are used to power the DBS satellites, but there are times when the satellite is in darkness. At these times, nickel-hydrogen batteries are used and then they are recharged by the solar panels. Over the years, the solar panels are hit by particles in space and the batteries lose efficiency, which is the main reason that the satellite becomes inoperative. The DBS satellite transmits compressed digital video signals, which produces very high-definition picture quality.

# DBS Satellite Overview

All communication services, from military, police, radio and television, and even communications satellites are assigned special bands of frequencies in a certain electromagnetic spectrum in which they are to operate.

To receive signals from the earth and relay them back again, satellites use very high frequency radio waves that operate in the microwave frequency bands. These are referred to as the *C band* or *KU band*. C-band satellites generally transmit in the frequency band of 3.7 to 4.2 Gigahertz (GHz), and is called the *Fixed Satellite Service band (FSS)*. However, these are the same frequencies occupied by ground-based point-to-point communications, making C-band satellite reception more susceptible to various types of interference.

The KU-band satellites are classified into two groups. The first include the low- and medium-power KU-band satellites, transmitting signals in the 11.7- to 12.2-GHz FSS band. And the new high-power KU-band satellites that transmit in the 12.2-GHz to 12.7-GHz Direct Broadcast Satellite service (DBS) band.

Unlike the C-band satellites, these newer KU-band DBS satellites have exclusive rights to the frequencies they use, and therefore have no microwave interference problems. The RCA system receives programming from high-power KU-band satellites operating in the DBS band.

The C-band satellites are spaced closed together at locations of 2 degrees. The high-power KU-band DBS satellites are spaced at 9 degrees, with a transmitter power of 120 watts or more.

Because of their lower frequency and transmitting power, C-band satellites require a larger receiving dish, anywhere from 6 to 10 feet in diameter. These whopper platters are at times referred to as "BUDs" or "Big Ugly Dish." The higher power of KU-band satellites enables them to broadcast to a compact 18-inch diameter dish.

# How the Satellite System Works

A satellite system is comprised of three basic elements:

**1** An uplink facility, which beams programming signal to satellites orbiting over the earth's equator at more than 22,000 miles.
**2** A satellite that receives the signals and retransmits them back down to earth.
**3** A receiving station, which includes the satellite dish. An RCA satellite receiver is shown in Fig. 6-2.

The picture and sound data information originating from a studio or broadcast facility is first sent to an uplink site, where it is processed and combined with other signals for transmission on microwave frequencies. Next, a large uplink dish concentrates these outgoing microwave signals and beams them up to a satellite located 22,247 miles above the equator. The satellite's receiving antenna captures the incoming signals and sends them to a receiver for further processing. These signals, which contain the original picture and sound information, are converted to another group of microwave frequencies, then sent up to an amplifier for transmission back to earth. This complete receiver/transmitter is called a *transponder*. The outgoing signals from the transponder are then reflected off a transmitting antenna, which focuses the microwaves into a beam of energy that is directed toward the earth. A satellite dish on the ground collects the microwave energy containing the original picture and sound information, and focuses that energy into a *low-noise block con-*

**FIGURE 6-2**    Front view of the DBS receiver.

*verter (LNB)*. The LNB amplifies and converts the microwave signals to yet another lower group of frequencies that can be sent via conventional coaxial cable to a satellite receiver-decoder inside your home. The receiver tunes each of the individual transponders and converts the original picture and sound information into video and audio signals that can be viewed and listened to on your conventional television receiver and stereo system.

## HOW THE RCA SYSTEM WORKS

The RCA DSS system is a DBS system. The complete system transports digital data, video, and audio to your home via high-powered KU-band satellites. The program provider beams its program information to an uplink site, where the signal is digitally encoded. The uplink site compresses the video and audio, encrypts the video and formats the information into data "packets." The signal is transmitted to DBS satellites orbiting thousands of miles above the equator at 101 degrees West longitude. The signal is then relayed back to earth and decoded by your DSS receiver system. The DSS receiver is connected to your phone line and communicates with the subscription service computer providing billing information on pay-per-view movies, etc. Figure 6-3 illustrates the overall operation of the DSS satellite system.

Now, here's a technical overview at how the total DSS system transports the digital signals from the ground stations via satellites into your home.

**Ground station uplink**  The program provider sends its program material to the uplink site, where the signal is then digitally encoded. The "uplink" is the portion of the signal transmitted from the earth to the satellite. The uplink site compresses the video and audio, encrypts the video, and formats the information into data "packets" that are then transmitted with large dishes up to the satellite. After this signal is received by the satellite, it is relayed back to earth and received by a small dish and decoded by your receiver.

**MPEG2 video compression**  The video and audio signals are transmitted as digital signals, instead of conventional analog signals. The amount of data required to code all of the video and audio information would require a transfer rate well into the hundreds of *Mbps (megabits per second)*. This would be too large and impractical a data rate to be processed in a cost-effective way with current equipment. To minimize the data-transfer rate, the data is compressed using *MPEG2 (Motion Picture Expert Group)*, a specification for transportation of moving images over communication data networks. Fundamentally, the system is based on the principle that images contain a lot of redundancy from one frame of video to another as the background stays the same for many frames at a time. Compression is accomplished by predicting motion that occurs from one frame of video to another and transmitting motion data and background information. By coding only the motion and background difference, instead of the entire frame of video information, the effective video data rate can be reduced from hundreds of Mbps to an average of 3 to 6 Mbps. This data rate is dynamic and will change, depending on the amount of motion occurring in the video picture.

In addition to MPEG video compression, MPEG audio compression is also used to reduce the audio data rate. Audio compression is accomplished by eliminating soft sounds that are near the loud sounds in the frequency domain. The compressed audio data rate can vary from 56 *Kbs (kilobits per second)* on mono signals to 384 Kbps on stereo signals.

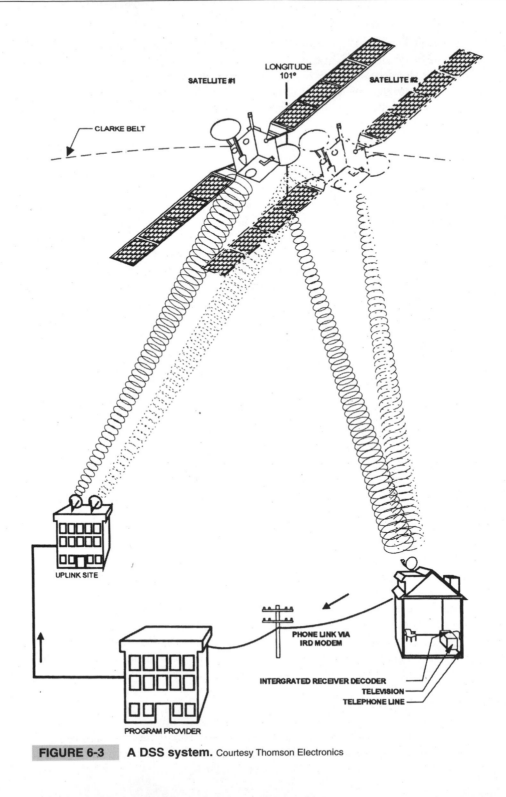

**FIGURE 6-3    A DSS system.** Courtesy Thomson Electronics

**Data encryption**  To prevent unauthorized signal reception, the video signal is encrypted (scrambled) at the uplink site. A secure encryption "algorithm" or formula, known as the *Digital Encryption Standard (DES)* is used to encode the video information. The keys for decoding the data are transmitted in the data packets. Your customer Access Card decrypts the keys, which allows your receiver to decode the data. When an Access Card is activated in a receiver for the first time, the serial number of the receiver is encoded on the Access Card. This prevents the Access Card from activating any other receiver, except the one in which it was initially authorized. The receiver will not function when the Access Card has been removed. At various times, the encryption will be changed and new cards will be issued to you to protect any unauthorized viewing.

**Digital data packets**  The video program information is completely digital and is transmitted in data "packets." This concept is very similar to data transferred by a computer over a modem. The five types of data packets used are Video, Audio, CA, PC compatible serial data, and Program Guide. The video and audio packets contain the visual and audio information of the program. The CA (Conditional Access) packet contains information that is addressed to each individual receiver. This includes customer E-Mail, Access Card activation information, and which channels the receiver is authorized to decode. PC compatible serial data packets can contain any form of data the program provider wants to transmit, such as stock market reports or software. The Program Guide maps the channel numbers to transponders and also gives you TV program listing information.

Figure 6-4 shows a typical uplink block diagram for one transponder. In the past, a single transponder was used for a single satellite channel. With digital signals, more than one satellite channel can be sent out over the same transponder. Figure 6-4 illustrates how one transponder handles three video channels, five stereo audio channels (one for each video channel plus two extra for other services, such as second language), and a PC-compatible data channel. Audio and video signals from the program provider are encoded and converted to data packets. The configurations can vary, depending on the type of programming to be put on stream. The data packets are then multiplexed into serial data and sent to the transmitter.

Each data packet contains 147 bytes. The first two bytes (remember, a byte consists of 8 bits) of information contained in the SCID (Service Channel ID). The SCID is a unique 12-bit number from 0 to 4095 that uniquely identifies the packet's data channel. The Flags consist of 4-bit numbers, used primarily to control whether or not the packet is encrypted and which key to use. The third byte of information is made up of a 4-bit Packet-Type indicator and a 4-bit Continuity Counter. The Packet Type identifies the packet as one of four data types. When combined with the SCID, the Packet Type determines how the packet is to be used. The Continuity Counter increments once for each Packet Type and SCID. The next 127 bytes of information consists of the "payload" data, which is the actual usable information sent from the program provider. The complete Data Packet is illustrated in Fig. 6-5.

## THE DIRECTTV SATELLITES

Two high-power KU-band satellites provide the DBS signal for the receiver. The satellites are located in a geostationary orbit in the Clarke belt, more than 22,000 miles above the

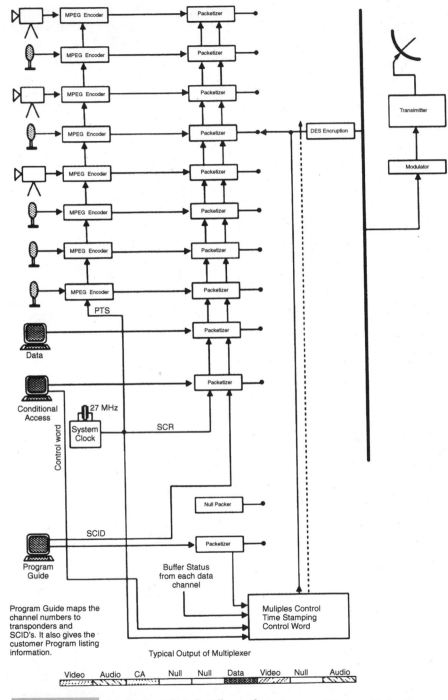

**FIGURE 6-4** **Typical up-link configuration.** Courtesy of Thomson Electronics

2 bytes    1 bytes

| | | 127 Bytes | 17 Bytes |

payload                Forward error correction

SCID & flags .        Packet type &
                      continuity counter

**FIGURE 6-5    An illustration of the data packets.** Courtesy Thomson Electronics

Right hand circularly polarized wave          Left hand circularly polarized wave

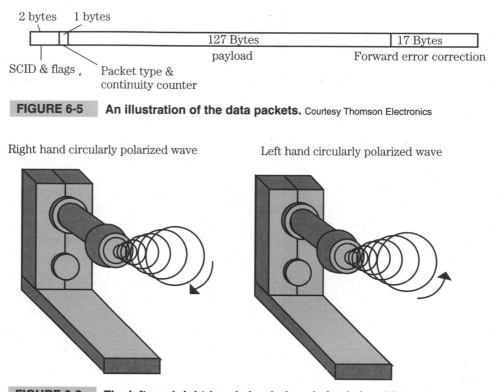

**FIGURE 6-6    The left- and right-hand circularly polarized signal from the satellites.** Courtesy Thomson Electronics

equator. They are positioned less than ½ degrees apart from each other with the center between them at 101 degrees West Longitude. This permits a fixed antenna to be pointed at the 101-degree slot and you are able to receive signals from both satellites. The downlink frequency is in the K4 part of the KU-band at a bandwidth of 12.2 GHz to 12.7 GHz. The total transponder channel frequency bandwidth is 24 MHz per channel, with the channel spacing at 14.58 MHz. Each satellite has 16 different 120-watt transponders. The satellites are designed to have a life expectancy of 12 or more years.

Unlike C-band satellites that use horizontal and vertical polarization, the DBS satellites use circular polarization. The microwave energy is transmitted in a spiral-like pattern. The direction of rotation determines the type of circular polarization (Fig. 6-6). In the DBS system, one satellite is configured for only right-hand circular-polarized transponders and the other one is configured for only left-hand circular polarized transponders. This results in a total of 32 transponders between the two satellites.

Although each satellite has only 16 transponders, the channel capabilities are far greater. Using data compression and multiplexing, the two satellites working together have the possibility of carrying over 150 conventional (non-HDTV) audio and video channels via 32 transponders.

**FIGURE 6-7**    **A roof-mounted DBS dish being installed and adjusted.**

**Dish operation**  The "dish" is an 18-inch, slightly oval-shaped KU-band antenna. The slight oval shape is caused by the 22.5-degree offset feed of the LNB (Low Noise Block converter), which is depicted in Fig. 6-7. The offset feed positions the LNB out of the way so that it does not block any surface area of the dish, preventing attenuation of the incoming microwave signal. Figure 6-7 shows the DBS dish being installed on a roof.

**Low-noise block (LNB)**  The LNB converts the 12.2-GHz to 12.7-GHz downlink signal from the satellites to the 950-MHz signal required by the receiver. Two types of LNBs are available: dual and single output. The single-output LNB has only one RF connector, but the dual-output LNB has two (Fig. 6-8). The dual-output LNB can be used to feed a second receiver or other form of distribution system. Figure 6-9 illustrates how the signal path is received from the satellite. The basic package comes with a single-output LNB. The deluxe receiver system has the dual-output LNB installed in the dish.

Both types of LNBs can receive both left and right-hand polarized signals. Polarization is selected electrically with a dc voltage fed onto the center connector and shield of the coax cable from the receiver. The right-hand polarization is selected with +13 volts while the lefthand polarization mode is selected with +17 Vdc. If you suspect coax or connector trouble, you can check for this dc voltage at the dish and at the antenna terminal on the back of the receiver. If you have proper dc voltage at the receiver antenna connection, but very low or no voltage at the dish coax connection, the cable is bad and needs to be replaced. Use a volt/ohmmeter for this check.

**Receiver circuit operation**    The DBS receiver is a very complex digital signal processor. The amount of speed and data that the receiver processes rivals even the fastest personal computers (PCs) on the market at this time. The information received from the satellite is a digital signal that is decoded and digitally processed. No analog signals are found, except for those exiting the NTSC video encoder and the audio *DAC (digital-to-analog converter)*. A block diagram of the DBS receiver is shown in Fig. 6-10.

The downlink signal from the satellite is downconverted from the 12.7- to 12.2-GHz range to the 950- to 1450-MHz range by the LNB converter. The tuner then isolates a single digitally modulated 24-MHz transponder. The demodulator converts the modulated data to a digital data stream.

The data is encoded at the transmitter site by a process that enables the decoder to reassemble the data and verify and correct errors that might have occurred during the transmission. This process is called *forward error correction (FEC)*. The error-corrected data is output to the transport IC via an 8-bit parallel interface.

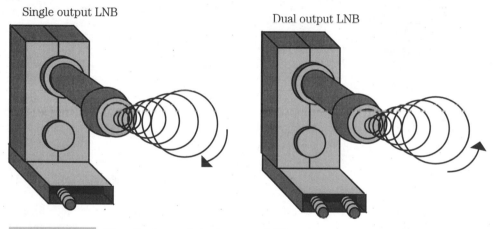

Single output LNB

Dual output LNB

**FIGURE 6-8**    **The single- and dual-output LNB.** Courtesy Thomson Electronics

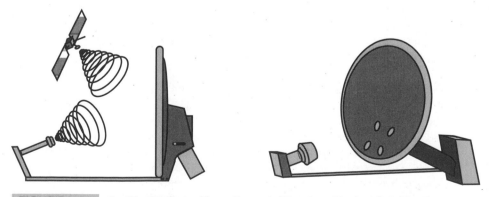

**FIGURE 6-9**    **An illustration of how the satellite signal is received by the antenna.** Courtesy Thomson Electronics

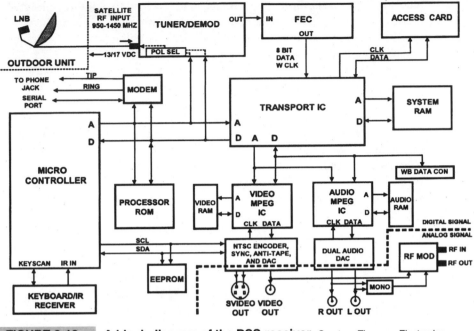

**FIGURE 6-10**   **A block diagram of the DSS receiver.** Courtesy Thomson Electronics

The transport IC is the heart of the receiver data-processing circuitry. Data from the FEC block is processed by the transport IC and sent to respective audio and video decoders. The microprocessor communicates with the audio and video decoders through the transport IC. The Access-Card interface is also processed through the transport IC and is used to turn on or validate the receiver.

The Access Card receives the encrypted keys for decoding a scrambled channel from the transport IC. The Access Card decrypts the keys and stores them in a register in the transport IC. The transport IC uses the keys to decode the data. The Access Card also handles the tracking and billing for these services.

Video data is processed by the MPEG video decoder. This IC decodes the compressed video data and sends it to the NTSC encoder. The encoder converts the digital video information into NTSC analog video that is then made available to the S-Video and the standard composite video output jacks.

Audio data is also decoded by the MPEG audio decoder. The decoded 16-bit stereo audio data is sent to the dual DAC, where the left and right audio-channel data are separated and converted back into stereo analog audio. The audio is the fed to the left and right audio jacks and is also mixed together to provide a mono audio source for the RF converter.

The microprocessor receives and decodes IR remote commands and front-panel keyboard commands. Its program software is contained in the processor ROM (Read Only Memory). The microprocessor controls the other digital devices of the receiver via the address and data lines. It is responsible for turning on the green LED on the on/off button.

**The receiver modem**   The modem in the receiver connects to your phone line and calls the program provider and transmits the pay-per-view programs purchased and reports them for billing purposes. The modem operates at 1200 bps and is controlled by the microprocessor. When the modem first attempts to dial, it sends the first number as touch-tone. If the dial tone continues after the first number, the modems switches to pulse dialing and redials the entire number. If the dial tone stops after the first number, the modem continues to dial the rest of the number as a touch-tone number. The modem also automatically releases the phone line if you pick up another phone on the same home extension.

**Diagnostic test menus**    The DBS receiver contains two diagnostic test menus. The first test is a customer-controlled menu that checks the signal, tuning, phone connections and the access card. The second test menu is servicer controlled. It checks out the majority of the receiver for problems.

**Customer-controlled diagnostics**    The customer controlled test helps you, the customer, during installation or any time the receiver appears not to function properly.

- *Signal test*    Checks the value of error bit number and the error rate to determine if the antenna connections are properly installed.
- *Tuning test*    Checks to ensure that a transponder can be tuned. The test is considered successful and this part of the test is halted if proper tuning occurs on 1 of the 32 transponders.
- *Phone test*    The phone test checks for dial tone and performs an internal loopback test. In Fig. 6-11, the system test indicates a phone connection problem. Some checks you can make is to plug a working phone into the phone line plugged into the back of the DBS receiver. If the phone works OK, the receiver modem could be defective. You will

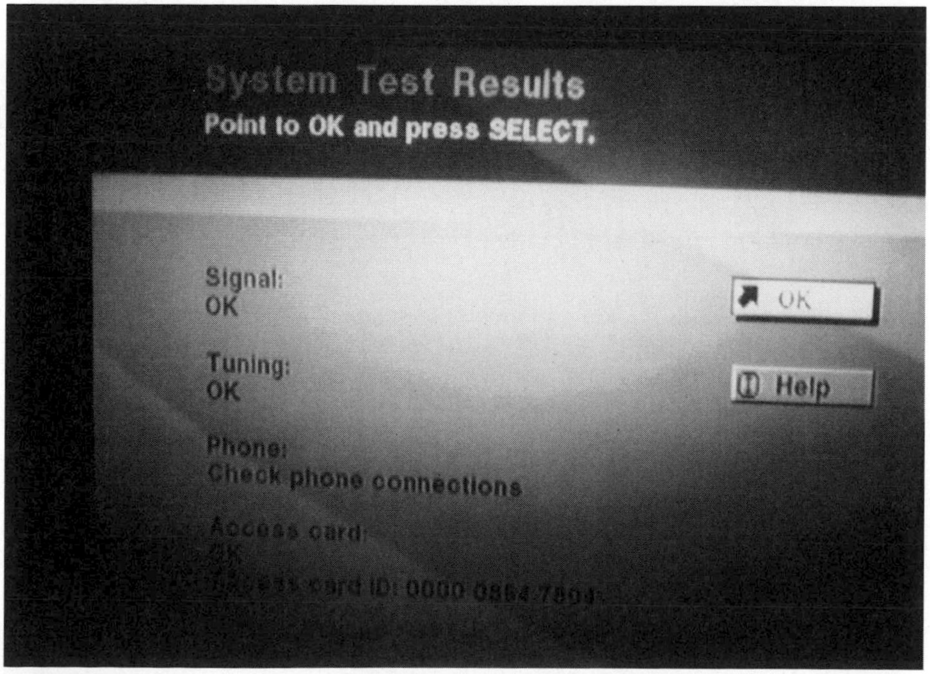

**FIGURE 6-11**    How the system test results look on your TV.

need to take it to a repair station for service. If the test phone does not work, check all plug-in connections or replace the phone cable to the unit.

■ *Access card test*  This test sends a message to the Access Card and checks for a valid reply.

The response for all tests will be an "OK" display or an appropriate message informing you of the general nature of the problem.

To enter the system test feature:

■ Select "Options" from the "DBS Main Menu." See Fig. 6-12.
■ Next, select "Setup" from the "Options" menu (Fig. 6-13).
■ Now select "System Test" from the "Setup" menu. Your TV screen will appear as in Fig. 6-14. Next, select "Test" from the System Test menu (Fig. 6-15).

---

DSS Main Menu
Use arrows to point to an item, then press SELECT.
Pressing the number also selects the number.

    1 Program guide
    2 Attractions
    3 Mailbox
    4 Options
    5 Alternate audio
    6 Help
    0 Exit

    4 Review purchases and set up your DSS system.

**FIGURE 6-12**    **The DSS main menu that is now in the**
**Options mode.** Courtesy Thomson Electronics

---

Options
Use arrows to point to an item, then press SELECT.
Pressing the number also selects the number.

    1 Past purchases
    2 Upcoming purchases
    3 Locks, limits, & channel lists
    4 Select channel list
    5 Set up
    0 Exit

    5 Set up your DSS system.

**FIGURE 6-13**    **The options menu in the "Set-up"**
**mode.** Courtesy Thomson Electronics

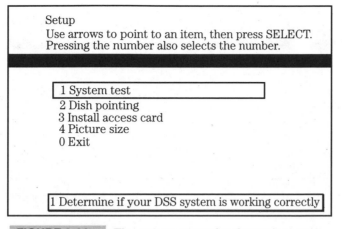

**FIGURE 6-14**    The set-up menu for the system test.
Courtesy Thomson Electronics

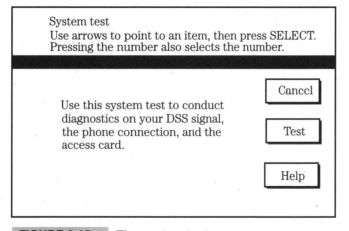

**FIGURE 6-15**    The system test menu.
Courtesy Thomson Electronics

The system test results are displayed automatically when the test is completed. The following two screens (Fig. 6-16) show whether the receiver passed or failed the test. If the Access Card passes the test, the Access Card ID number will be displayed in the window of the menu.

# Controlled Diagnostics
# for Troubleshooting

The servicer-controlled test provides a more in-depth analysis of the receiver and overall system operation for proper operation. The test pattern checks all possible connections

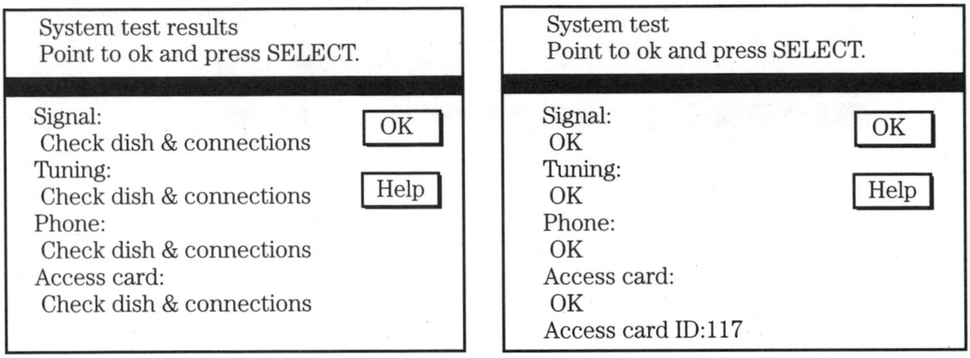

```
┌──────────────────────────────┐   ┌──────────────────────────────┐
│ System test results          │   │ System test                  │
│ Point to ok and press SELECT.│   │ Point to ok and press SELECT.│
├──────────────────────────────┤   ├──────────────────────────────┤
│ Signal:                ┌────┐ │   │ Signal:                ┌────┐ │
│   Check dish & connections│OK│ │   │   OK                  │ OK │ │
│ Tuning:                └────┘ │   │ Tuning:                └────┘ │
│   Check dish & connections┌────┐│   │   OK                  ┌────┐ │
│ Phone:                 │Help│ │   │ Phone:                │Help│ │
│   Check dish & connections└────┘│   │   OK                  └────┘ │
│ Access card:                  │   │ Access card:                 │
│   Check dish & connections    │   │   OK                         │
│                               │   │ Access card ID:117           │
└──────────────────────────────┘   └──────────────────────────────┘
```

**FIGURE 6-16**    **The menu for the System Test Results.** Courtesy Thomson Electronics

between components as a troubleshooting aid. The following information is provided for system diagnostics.

**1** IRD (receiver) serial number
**2** Demodulator vendor and version number
**3** Signal strength
**4** ROM checksum results
**5** SRAM test results
**6** VRAM test results
**7** Telco (phone) callback results
**8** Verifier Version
**9** Access Card Test and Serial Number
**10** IRD ROM version
**11** EEPROM test results

The response for all of these tests will indicate the test was or was not successful.

In addition, this menu will allow entry into the phone prefix menu so that the installer can set up a one-digit phone prefix.


## SERVICE TEST

To enter the service test mode feature of the DBS system, use the front-panel buttons of the receiver, not the remote control unit. For service test, simultaneously press the front-panel "TV/DBS" and the "Down" arrow buttons. The following screen menus will come up on your TV (Fig. 6-17).

The test results are automatically displayed after the test is completed. You or the service technician are given the option to exit the test or run the diagnosis again.

**Front-panel control buttons**  Also included in the Service Test Menu are provisions for testing the modem and setting a single-digit phone number. During the service test, the modem will dial the phone number that appears in the boxes at the top of the test menu. The phone number can be changed by using the "Down" arrow keys on the remote control or receiver to move the cursor past the "Prefix" prompt to the number boxes. Once the boxes are selected, the number can be entered or changed with the number keys on the remote or by using the "Up"/"Down" keys on the remote or the receiver. The prefix can be changed by selecting the "Phone Prefix" on the display and changing the number with the number keys on the remote control or by using the arrow keys on the remote-control hand unit or the receiver front panel. The receiver front-panel control buttons are shown in Fig. 6-18.

**Pointing the dish**  When you are installing your dish, you have to consider where to locate the dish so as not to have any trees or buildings blocking the signal from the satellite. Figure 6-19 shows the dish pointing from sky-high view. You first have to determine the satellite's position in the sky. You determine the side to side (azimuth) and the up/down (elevation) bearings from your location to the satellite. These change with different locations across the United States. For example, the azimuth and elevation for Minneapolis are different from those in Houston, Texas. These changes are caused by the satellite's position in the geostationary orbit. Also, the azimuth of the dish changes as you move either east or west.

Service test

| IRO | #01020304FF |
| Demod type | HNS |
| Demod signal | 100,-158 |
| ROM IC3151 | 55 |
| SRAM IC3351 | OK |
| V-Drum test | OK |
| Telco IC3201 | No dial tone |
| Verifier cam | 10A,00   ROM version 0750 |
| Access card | 117   EEprom IC3161  OK |

Test
Stop
Phone prefix

Phone prefix
Enter the phone prefix, then
Point to ok and press SELECT.

Prefix:

OK
None
Help

Enter 1-digit phone prefix to get a local outside line if necessary.

**FIGURE 6-17**    **The Service Test menus.** Courtesy Thomson Electronics

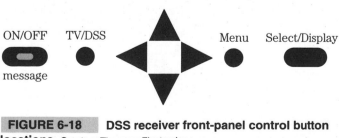

ON/OFF    TV/DSS              Menu    Select/Display

message

**FIGURE 6-18**    **DSS receiver front-panel control button locations.** Courtesy Thomson Electronics

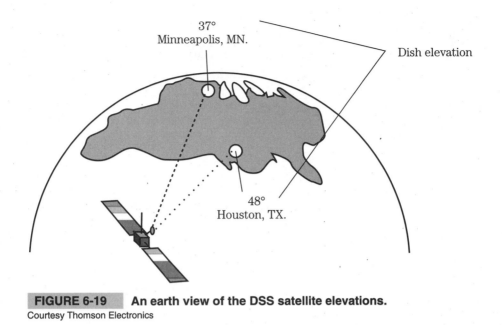

**FIGURE 6-19**    **An earth view of the DSS satellite elevations.**
Courtesy Thomson Electronics

# A World View of the DSS System

One way to better understand the DBS system is to look at the different parts of the system—from the studio down to the DSS receiver and the remote-control unit that you are using in your home. Refer to the "bird's eye" view of the DSS system (Fig. 6-20).

- *Uplink control center*  This building houses the equipment that transmits the programming via very large dishes up to the orbiting satellites.
- *Satellite*  The satellite relays the programming signals back to your satellite dish. The satellite is parked above the equator, in a geostationary orbit 22,300 miles above the earth.
- *Dish antenna*  The small dish receives the satellite signals. Because the satellite is so powerful, the dish only needs to have a diameter of 18 inches.

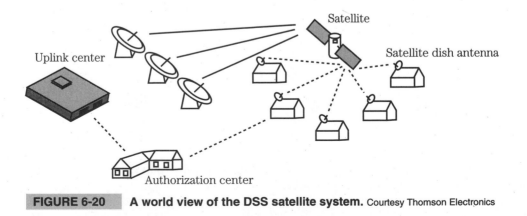

**FIGURE 6-20**    **A world view of the DSS satellite system.** Courtesy Thomson Electronics

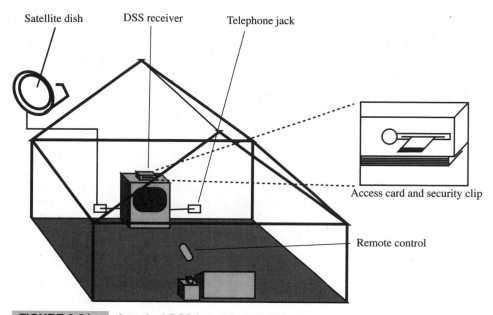

Satellite dish     DSS receiver     Telephone jack

Access card and security clip

Remote control

**FIGURE 6-21**     **A typical DSS home installation set-up.** Courtesy Thomson Electronics

■ *Program provider authorization center*  This center processes your billing statements. Your system is linked to the customer service center through the phone jack located on the back of your receiver.

■ *DSS system home view*  Figure 6-21 illustrates the parts of the system required inside your home and satellite dish outside the house.

■ *Receiver*  The receiver receives the TV digital program information and sends it to your TV for viewing or to a VCR for recording.

■ *Telephone jack*  A cable from this jack connects to back of the receiver and is plugged into the wall phone jack. The receiver uses a toll-free number once per month to update your Access Card. This update only takes a few seconds and ensures that you will have continuous service. The system automatically hangs up if you pick up the phone when the receiver is calling out the update information.

■ *Television*  If your TV is remote controllable, you can program the Universal TV remote to change channels and the volume level.

■ *TV universal remote*  The Universal remote is included with the receiver. This unit not only controls the system, but most other remote controllable TVs. Just point the remote at the set you want to control.

■ *Access card*  Each receiver must have an Access Card. The card must be inserted before you can use the system. The card provides system security and authorization of DSS services. Do not remove the card, except when issued a new card as a replacement for the original.

■ *Security card clip*  The clip is installed (Fig. 6-21). This clip fits over the Access Card and helps prevent the card from being inadvertently removed. To remove the clip, squeeze the top and bottom together and slide the clip off of the Access Card.

## FRONT-PANEL RECEIVER CONTROLS

The following is information on front-panel receiver operation and functions.

■ *On/off/message control*   This control turns the receiver on and off. When the receiver is turned off, a flashing light indicates that a message has been sent by the customer service center. The receiver never actually has the power turned off, but is put into a Standby mode.

■ *TV/DSS switch*   This button switches the "OUT TO TV" connection from DSS programming to the normal TV antenna or cable input. This is similar to the TV/VCR button on many VCRs.

■ *Arrow keys*   These keys allow you to move around the program guide and menu to make your selections. Use these arrows to point before selecting an item on the menu. When you are not in the program guide or the menu system, the up/down arrows can be used to change channels.

■ *Menu button*   This button brings up the menu on the screen of your TV for program selections.

■ *Select/display button*   Push this button to select an item you have pointed to when using the program guide or menus. Also brings up a channel marker showing the time, channel, and other program details when you are viewing a program or previewing a coming attraction.

The front-panel receiver buttons can be used to control the receiver when the remote control is not close by. See Fig. 6-22 for these front panel control locations.

■ *Access Card slot*   Insert the Access Card in the receiver with the arrow face up and pointing toward the unit. The receiver is shipped with the Access Card inserted into the slot. Do not remove the Access Card, except to install a new card issued as a replacement for the original card.

Do not stack electronic components or other objects on top of the receiver. The slots on top of the receiver must be left uncovered to allow for proper airflow circulation to the unit. Blocking airflow to the receiver could degrade performance or cause damage to your receiver or to other components.

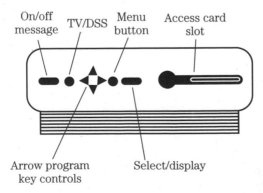

**FIGURE 6-22**   **Call-outs of a typical DBS receiver control locations.**

# Connecting the Receiver

The following four drawings give you some examples of some hookups that are generally used to connect the receiver to a TV or other components. You can also refer to your TV and VCR owner's manuals for more specific information in regards to connecting your own components.

## CONNECTION A

This hookup (Fig. 6-23) provides the best possible picture and stereo audio quality. To use connection A, you must have:

■ TV with S-Video input, plus separate RF and audio/video inputs (jacks).
■ VCR with RF input and output.
■ S-Video, coaxial, and audio/video cables.

## CONNECTION B

This hookup (Fig. 6-24) provides good picture and stereo audio quality. To use connection B, you must have:

■ TV with separate RF and audio/video inputs (jacks).
■ VCR with RF input and output.
■ Coaxial and audio/video cables.

## CONNECTION C

This hookup (Fig. 6-25) provides good picture and mono audio quality. To use connection C, you must have:

■ TV with RF input (jack).
■ VCR with RF input and output.
■ Coaxial cables.

## CONNECTION D

These connections (Fig. 6-26) provide a good picture and mono audio quality. To use connection D, you must have the following items:

■ TV with an RF input (jack).
■ Coaxial cables for the connections.

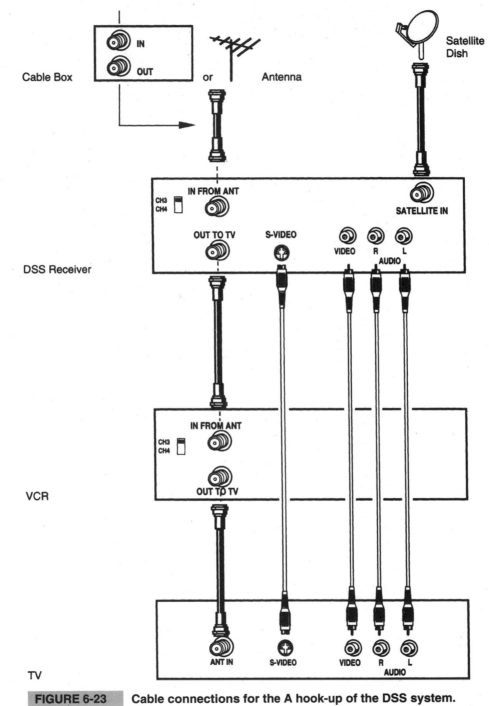

Cable Box

IN

OUT

or

Antenna

Satellite Dish

DSS Receiver

CH3
CH4

IN FROM ANT

OUT TO TV

S-VIDEO

VIDEO

R

L

AUDIO

SATELLITE IN

VCR

CH3
CH4

IN FROM ANT

OUT TO TV

TV

ANT IN

S-VIDEO

VIDEO

R

L

AUDIO

**FIGURE 6-23** **Cable connections for the A hook-up of the DSS system.**
Courtesy Thomson Electronics

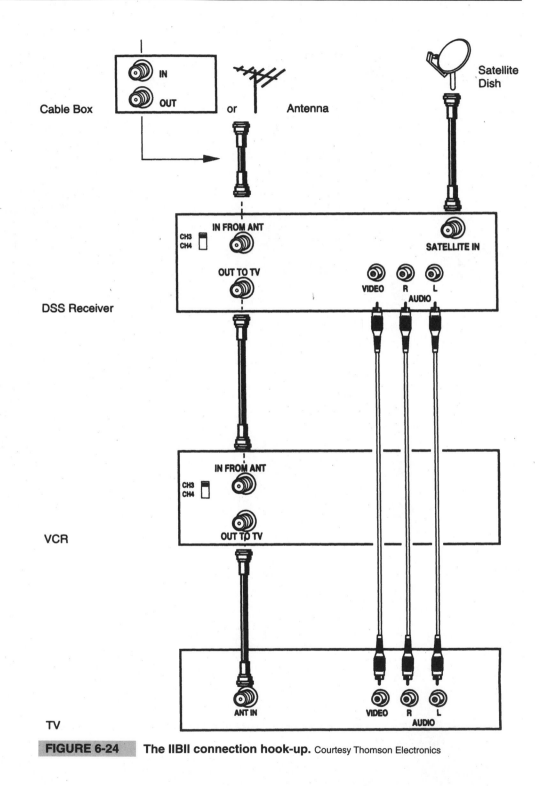

**FIGURE 6-24**    **The IIBII connection hook-up.** Courtesy Thomson Electronics

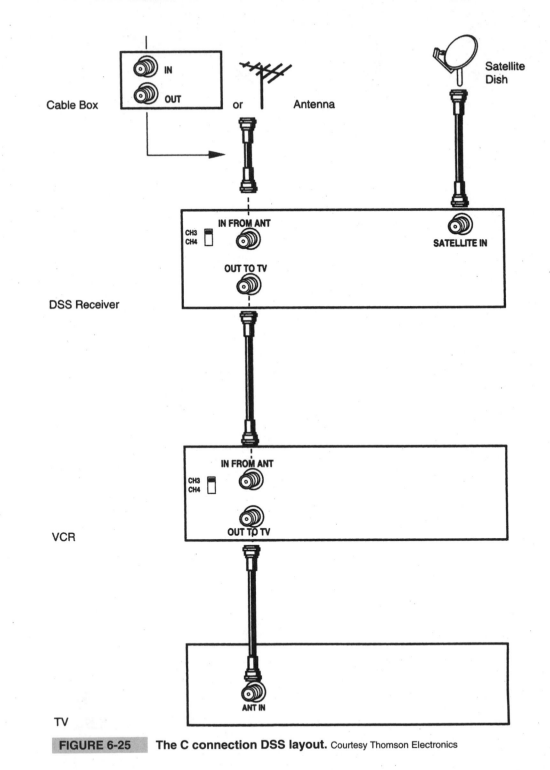

**FIGURE 6-25** **The C connection DSS layout.** Courtesy Thomson Electronics

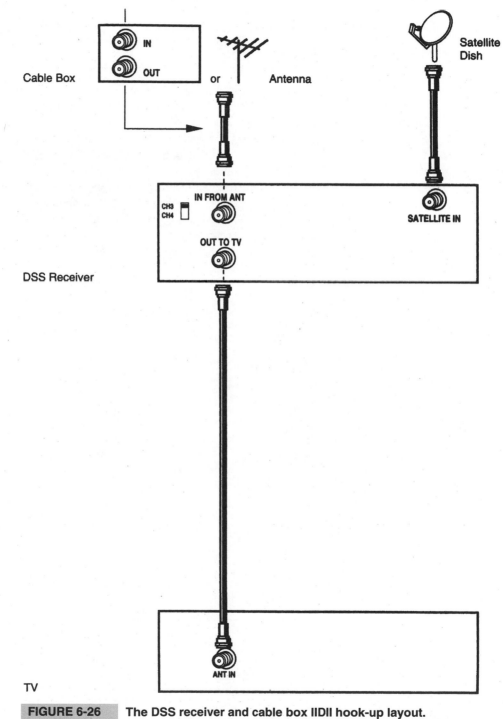

Cable Box

IN

OUT

or    Antenna

Satellite
Dish

IN FROM ANT

CH3
CH4

OUT TO TV

SATELLITE IN

DSS Receiver

ANT IN

TV

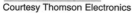 **FIGURE 6-26**    **The DSS receiver and cable box IIDII hook-up layout.**
Courtesy Thomson Electronics

# Some Possible DBS System Problems and Solutions

You might lose your DBS signal if the dish becomes covered in snow (Fig. 6-27) or frozen with ice. Clean the snow off or melt the ice. Also, some dishes have a built-in de-icer for cold weather conditions and some heater kits can also be installed. The ice or snow on the dish cuts the microwave signal down very low and the TV will display a "hunting for signal" message. Also, recheck the position of the dish because it might have been turned a bit from ice and windy winter storm conditions.

Other loss-of-signal conditions or intermittent receiver operation could be due to loose cable connections or defective coax cables. Check the cables for an open center lead or shield or a short between the cable center lead and the outside shield. This can be checked with a very inexpensive ohmmeter. Refer to Chapter 1 of how to use the ohm and volt-meters. Be sure that all cable connections are crimped tight. Sometimes the center copper

**FIGURE 6-27**    Snow or ice build-up on the dish can prevent DBS reception. A dish heater is available that will melt any ice or snow.

connector lead wire will bend or will be too short to make a good contact. As explained earlier, just follow the TV screen troubleshooting menu.

For remote-control problems, check out or replace the batteries. Be sure that the battery contacts are clean and tight. If you see any green looking corrosion on the battery or the battery contacts, clean them so that they are clean and bright. Should the remote unit become wet or been dropped in water, remove the batteries and wipe dry as soon as possible. Then use a hair dryer to completely dry it out. If you can remove the case, you can dry it much easier with the hair dryer. After the remote is dried out, replace the batteries. There's a good chance that the remote will operate.

# DBS Glossary

*Access Card*   Identifies the DBS service providers and is required for your DSS system to work. Do not remove the Access Card, except when a new card has been issued to replace the original one.

*Alternate audio*   Refers to the different audio channels that can be broadcast in conjunction with a video program. A foreign-language translation is an example.

*Attractions*   Previews of special programs broadcast by your program provider.

*Azimuth*   Refers to the left-to-right positioning of your dish antenna. When you enter your zip code (or latitude and longitude), the display screen provides the number corresponding to an azimuth setting for your location.

*Channel limit*   Allows you to select which channels that can be viewed when the system is locked.

*Receiver*   Receives, processes, and converts the digitally compressed satellite signals into audio and video.

*Elevation*   Refers to the up and down positioning of your DBS dish. When you enter your zip code (or latitude and longitude), the display screen provides the number corresponding to the elevation setting for your location.

*Key*   The user-defined four-digit password that allows you to limit access to certain features of your DBS system.

*Limits*   There are three kinds of limits. The *Ratings Limit* allows you to control program viewing of rated programs by ratings level. The *Spending Limit* controls spending on a cost-per-program basis. The *Channel Limit* allows you to select which channel can be viewed when the system is locked.

*Locks*   The locks are a means of restricting access to certain features of the DBS system. The lock is controlled by a four-digit key that acts as a password. The closed or open lock icon in the channel marker indicates whether your system is locked or is unlocked.

*Mailbox*   Stores incoming electronic messages sent to you by your program providers. The mailbox is accessed through the on-screen menu system, and can store as many as 10 messages of 40 characters each.

*Main menu*   This is the first list of choices in the on-screen menu system. Press the Menu button on the remote-control unit or buttons on the receiver front panel to bring up the Main menu.

Some information in this chapter is courtesy of Thomson Consumer Electronics Company (RCA).

# HOW VIDEO CAMERAS
# AND CAMCORDERS WORK

# Camcorder Features and Selections

Now, a large selection of video camcorders are available. The older, large camcorder (Fig. 7-1) takes the standard up to 6-hour recording tape. These are still very popular. The much lighter weight and easier-to-carry models are the 8-mm and VHS-C camcorders. The VHS-C small tape can be put into an adapter and played back on a standard VHS machine. The standard VHS and VHS-C camcorders are easier to use for playback.

If you have a 8-mm VCR, then you would want an 8-mm camcorder to play back your tapes. One advantage of an 8-mm camcorder (Fig. 7-2) is that you can place two hours of recording on one 8-mm tape. With a VHS-C cassette camcorder, you have only 30 minutes (or up to 90 minutes with a much poorer quality).

**FIGURE 7-1**    A full-size VHS Zenith camcorder.

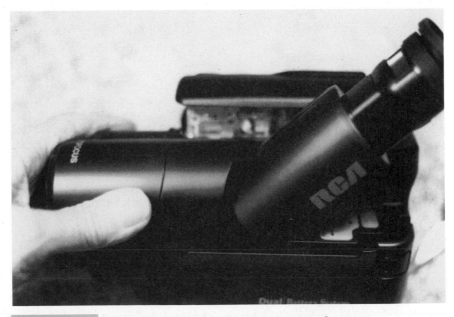

**FIGURE 7-2**    An 8-mm dual-battery system camcorder.

The JVC model GR-AX900 is VHS-C and it has the following features. It does nice time-lapse video, creates still shots and can then mix these videos into programs for editing functions when copying a tape to your VCR. The Sony model CCD-TRV40 has stabilization and a large zoom lens. The Sony also has a built-in speaker and a 3-inch LCD color monitor screen. This is great for reviewing the videos you just recorded and it can be used as a large viewfinder for recording.

# Digital Video Images

Digital TV video is now on the horizon and digital camcorder technology has been used for a few years. Camcorder digital techniques lets you do special effects, such as to merge one shot into another scene, enlarge a picture, or make still frames very sharp and jitter free. Camcorders now use computer-type digital coding with images consisting of zeros and ones. This also allows you to put your video pictures into your computer and perform all sorts of picture manipulations.

You will find that digital camcorders have the same size of cassettes, which eliminates different formats. The tiny MiniDV video cassette is smaller than an 8-mm or VHS-C cassette, but holds 60 minutes of recordings.

These tiny cassettes lets the camcorder companies build small, light-weight models that you can easily carry anywhere.

The digital Sharp model VL-DC1U is a little larger than other models in the View-Cam line and features a large color viewing screen of four inches.

The JVC model GR-DV1U weighs in around 18 ounces with the battery and tape installed. It is about the size of a paper back book and has great special recording effects.

The Sony digital camcorder weighs about 22 ounces and sports a 2½-inch color "pop out" viewfinder screen and has a built-in speaker.

To playback a digital tape, you will need to connect a cable from your camcorder and plug it into a TV. Another way is to connect the digital camcorder to a VCR to make an analog video tape, and then play it back through your TV. Digital TVs and recorders are scheduled for the marketplace in 1998.

# Video Camera/Camcorder Basics

There are actually only two types of video cameras, which are determined by the type of pick-up device they have to convert light to electronic signals. One camera type uses a vacuum-tube pick-up device and the other uses a solid-state pickup. The two types of solid-state pick ups are *CCD (charge-coupled devices)* and *MOS (metal-oxide semiconductor)*, although the CCD is more popular. More on the CCD chip later in this chapter. In the mid 1980s, all video cameras used the tube-type pick-up device for imaging. As solid-state imaging chips become available, they were quickly used in portable consumer cameras because of their small size and weight advantages.

As the cost of solid-state chips have decreased and their resolution and light sensitivity has increased, consumer, industrial, and even TV broadcasters now use solid-state CCD pick-up for cameras and camcorders. Other advantages of CCD pick-ups are their increased ruggedness, decreased image lag, better sensitivity, less power-consuming drive circuits, higher-level output signals, and a lot less circuitry that a vacuum pick-up tube requires.

# How Video Cameras Work

It really does not matter what type of pick-up device is used, the operating circuits are very similar from one video camera to another. Figure 7-3 shows the major circuits and signal flow of most basic camera types. Besides the lens and pick-up device, the signal processing and control circuits are very similar to circuits found in many other video products covered in this book.

## CAMERA TYPES

The video cameras are usually of two types. The camera will be of the stand-alone type that can be used with a monitor or tape recorder or a camera that is part of a camcorder.

Cameras first appeared for monitoring, and surveillance in conjunction with time-lapsed video tape recorders for security and military work. Some consumer cameras were also sold in the early 1980s. TV stations use stand-alone video cameras for programs and recording.

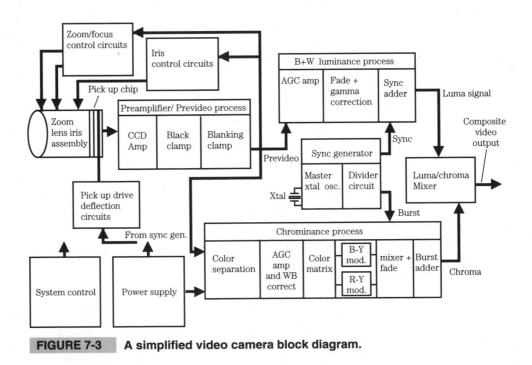

**FIGURE 7-3**    A simplified video camera block diagram.

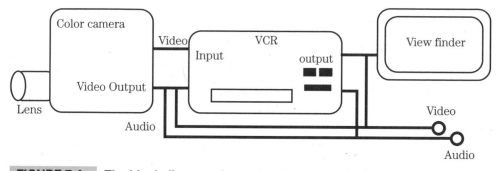

**FIGURE 7-4**    The block diagram shows that the camcorder is a combination of a camera, VCR, and viewfinder.

## WHAT IS A CAMCORDER?

Camcorders are in demand for consumers and for TV stations and networks for outside news gathering, etc. Camcorders combine a camera, a VCR record/playback section, and a viewfinder that is also used for looking at the video playback (Fig. 7-4). The camera in a camcorder also share some of its electronics, such as the power supply, control system, and video circuits, with the VCR portion.

**Usual camcorder faults**    The most common failures with a camcorder are caused by the mechanical nature of the VCR transport section and the camera lens, plus the handheld, portable nature for the way the camcorder is used. The same mechanical failures that occur in standard home VCRs also occur in the VCR section of camcorders, but usually not as often because they are not used as much. Worn rubber and broken gears are common failures with camcorders. The camera lens assembly, including the iris, focus, and zoom control motors and gears also have a high failure rate.

The lens problems, as well as broken circuit boards and poor/broken solder connections are usually caused by rough handling and dropping, which will occur with a handheld portable device. If you drop the camcorder on its lens, it could cause lens damage, motor problems, or stripped gears. These type of mechanical failures with the camera section are usually quite easy to diagnose and you may be able to repair your self.

Other portions of the camcorder that develops troubles in either the camera or VCR section is of an electronics nature. In the VCR section, this would be servo, cylinder head, pre-amp, chroma, luminance or black-and-white, power supply, and system-control stages. In the camera section, electronic failures include sync generator, CCD imager, chrome, luma, power supply, and control problems.

**Determining which camcorder section is faulty**    Now look at ways to localize camcorder problems:

■ *Localizing the problem area*    You need to determine whether the camcorder failure is related to the VCR, camera, or electronic viewfinder, and if the failure is mechanical or electrical. Then, see if you can correct the problem yourself or should you take the unit in for professional work.

- *Mechanical troubleshooting* Do this to isolate the worn or damaged mechanical parts, which are causing improper VCR or camera operation.
- *Electronic troubleshooting* This is performed to isolate the defective component that is causing the VCR or camera to operate incorrectly.
- *Alignment information* Use this to determine if your camcorder needs alignment or adjustments caused by wear, drift, normal usage, or parts that have been replaced. The alignment for camcorders requires special equipment, jigs, and technical skills.

**Performance check out** After you have performed any repairs or had your camcorder repaired at a service center, you should make some recordings and use all of the control functions to be sure that it is functioning properly.

**Video camera functional blocks** The following is a brief description of the operational blocks that make up a typical video camera. Notice that, depending on individual camera design, the layout order for some of the blocks might be a little different for various brands of cameras.

*Lens/iris/motors* The lens assembly focuses light from the scene you're viewing onto the light-sensitive surface of the pick-up device. The auto-iris circuit controls the amount of light that passes through the lens by operating a motor to open and close the iris diaphragm (Fig. 7-5). Under bright lighting conditions, the iris controls the amount of light falling on the pick-up device and thus the amplitude (strength) of the prevideo output signal. Proper operation of the auto-iris circuit is crucial for video output because the iris diaphragm is spring-loaded closed; a failure in the iris control or drive circuit prevents light from reaching the camera pick-up device. The focus drive circuit generates the signals necessary to operate the focus motor. In cameras with auto-focus, the control circuit reacts to high-frequency information in the prevideo signal, or to an infrared or LED sensor. The zoom drive circuit generates the signals necessary to operate the zoom motor by reacting to input from the camera zoom control-button contacts.

*Sync generator circuitry* The sync generator provides synchronization for all the other camera circuits. The output signals are developed by dividing down the signals from a

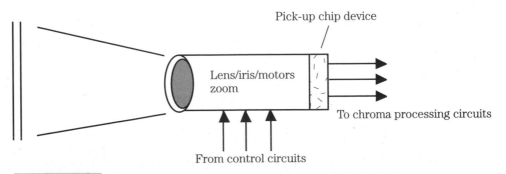

**FIGURE 7-5** The lens, iris, and control motors adjust the light that passes through to the pick-up device.

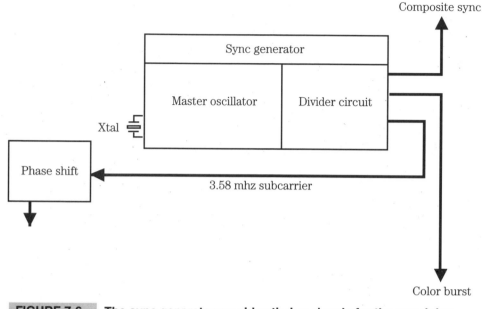

**FIGURE 7-6**  **The sync generator provides timing signals for the remaining camera stages.**

master crystal-controlled oscillator. The master oscillator typically operates at two, four, or eight times the 3.58-MHz chroma burst frequency. The sync generator provides horizontal and vertical drive signals to the pickup device, composite sync and burst for the video output, and 3.58-MHz subcarrier reference signals for the R-Y (red) and B-Y (blue) modulators. The block diagram for the sync generator operation is shown in Fig. 7-6.

*Camera pick-up devices*  Presently, three types of image pick-up devices are used in consumer, broadcast, and industrial video cameras. These are vacuum tube, MOS, and CCD (charge-coupled devices) devices. The CCD devices are solid-state pick-ups, made of a large number of photodiodes arranged horizontally and vertically in rows and columns (Fig. 7-7). CCDs are now the most commonly used image pick-up devices.

Tube pick-up devices (Vidicon, Saticon, and Newvicon are common types) use magnetic yoke deflection and a high-voltage supply to scan an electron beam across a light-sensitive surface. These tube pick-ups suffer the same scanning irregularities that television picture tubes have, plus more, and require many scan-correction circuits to produce an acceptable output signal. Also, the very low-level output signal from the tube pick-ups (200 µV or less) requires an extremely high gain, with a very low-noise preamplifier as the first signal stage. Tube pickups have been replaced by solid-state CCD and MOS image devices in consumer cameras/camcorders and are being phased out of most broadcast and industrial cameras.

Solid-state MOS and CCD pick-up devices are very similar to each other in operation and performance, with only a few significant differences. Conversion of light to electrical energy occurs at each of the individual photodiodes, which produce a small electrical

CCD drive/ pick-up chip

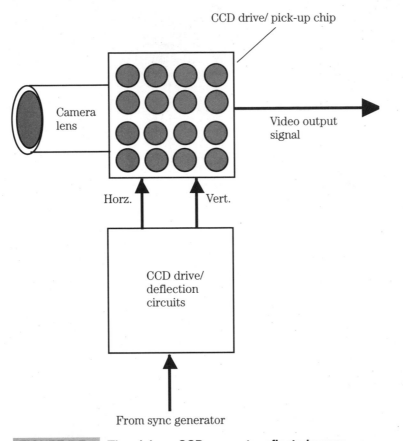

Camera lens

Video output signal

Horz.        Vert.

CCD drive/ deflection circuits

From sync generator

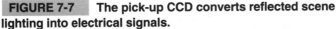

**FIGURE 7-7    The pick-up CCD converts reflected scene lighting into electrical signals.**

charge when light from the scene is focused on their exposed surface. A method of matrix scanning is used to repeatedly collect each of these charges and assemble them into a video signal. The scanning method used to collect these charges is one of the major differences between MOS and CCD pick-up devices.

MOS devices use a scanning method that results in three or four signal output lines. These lines carry white, yellow, cyan, and green color signals. (no green for the older three-line devices). One disadvantage of MOS devices is that the output signals are at a fairly low level (40 to 50 mV) and require low-noise preamps to bring the signals up to a usable level for standard signal color-processing circuits.

CCD devices use a scanning method that results in a single video output line. This signal contains all of the necessary luminance and chrominance information required to generate NTSC composite video. Also, the level of the output signal is high enough that no preamp is required. An advantage of CCD devices is that they have been more reliable than MOS devices.

With all types of pick-up devices, when color is desired, a multicolored filter is placed in front of the pick-up device's light-sensitive surface. This, along with the scanning of the device, results in the production of an extra high-frequency signal that carries information about color in the scene to be viewed. The location of the CCD chip is being pointed out in Fig. 7-8 photo of an 8-mm camcorder that has its side cover removed for service.

# Repairing and Cleaning Your Camcorder

This section shows how to repair and clean your camcorder. As you know, the camcorder is a combination of camera and small VCR. In fact, many of the service information tips found in the VCR chapter can be used for your camcorder.

The most popular camcorder sold today uses the 8-mm tape format. The cassette for 8 mm is thinner than the VHS-C and is about the size of an audio cassette. For this reason, the 8-mm camcorder can be made much smaller in size and thus much easier to carry around. The units usually weigh less than two pounds. Figure 7-9 shows a Zenith model VM8300 8-mm camcorder.

## TAKING YOUR CAMCORDER APART

To clean, repair, and make minor adjustments, you will have to take your camcorder apart. When you do, be very careful because very small and delicate parts are contained inside.

CCD Chip

**FIGURE 7-8**    **The CCD chip location in an 8-mm camcorder.**

**FIGURE 7-9**    **A Zenith model VM8300 8-mm camcorder.**

The screws are very small, so put them in a plastic cup or zip-top plastic bag, so as not to lose any of them. Also, remember where the screws come out of as they will be of different size and screw-thread types. You might want to draw a sketch as the case, screws, and parts are taken apart so that you will know how to put it back together after repairs and cleaning is completed.

**How to take apart the cassette lid and deck**    The following three drawings show all of the steps for taking apart a Zenith VHS-C camcorder. These same procedures can be used for most all models of camcorders.

**1** Refer to Fig. 7-10. Take out the two screws (A) that hold on the cassette cover. Raise the cassette cover, as indicated by the arrow (B) to remove. With this cover removed, you can usually clean the video head cylinder and other rubber roller parts and even part of the tape path.

**2** Take out the two screws (C) and remove the base assembly.

**3** Take out the three screws marked (D) and one screw (E).

**4** The front panel and side panel are engaged by a plastic rim. Carefully squeeze the portions of the side panel between your thumb and forefinger and raise the deck section slightly to disengage it.

**5** Disconnect the connectors (F), (G), (I), and (J). The deck and operation sections can now be separated from the camera section.

## Taking apart the lower case section

**1** Refer to Fig. 7-11. Take out the screws marked *A* and *B*, and remove the insulator sheet.
**2** Take out the screws (C), (D), and (E). Disengage the side panel from the lower case by shifting and raising it.

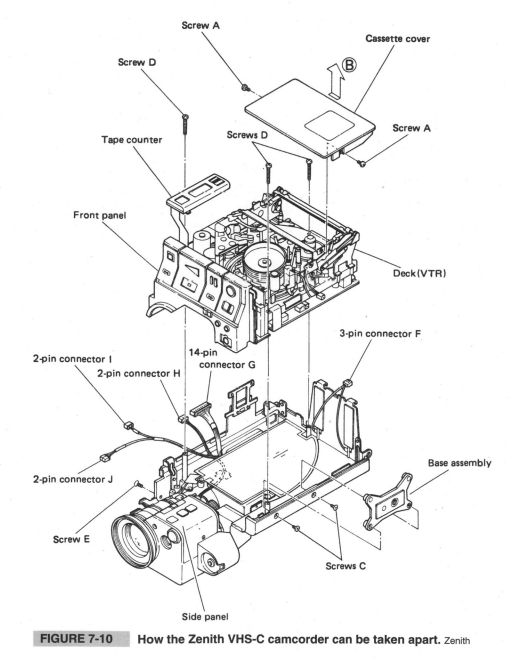

Screw A

Cassette cover

Screw D

Tape counter

Screws D

Screw A

Front panel

Deck (VTR)

3-pin connector F

2-pin connector I

14-pin connector G

2-pin connector H

2-pin connector J

Base assembly

Screw E

Screws C

Side panel

**FIGURE 7-10**     **How the Zenith VHS-C camcorder can be taken apart.** Zenith

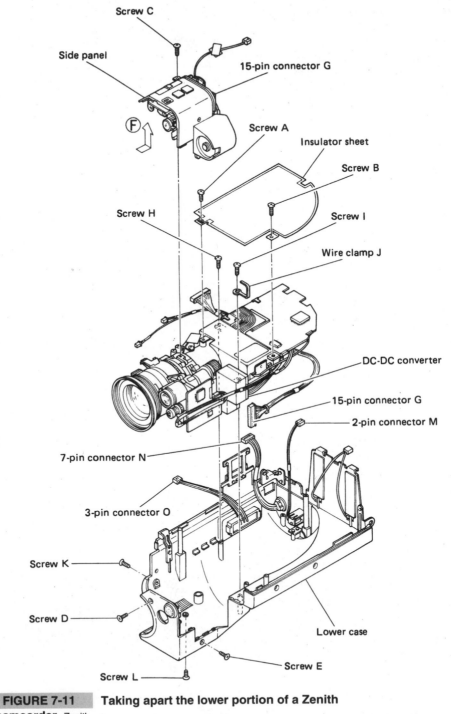

**FIGURE 7-11** **Taking apart the lower portion of a Zenith camcorder.** Zenith

**Taking apart the lower section of camcorder**  Now, see how to disassemble the lower part of the camcorder that contains the lens and camera section.

**1** Refer to Fig. 7-11. Take out screws labeled *A* and *B*, and remove the insulator sheet.
**2** Take out the screws (C), (D), and (E). Disengage the side panel from the lower case by shifting and raising it, as shown by the arrow labeled *F*. Disconnect the connector indicated by G.
**3** Remove screw H, screw I and wire clamp J. Then take out screws K and L.
**4** Raise the camera section slightly and disconnect connectors M, N, and O, which are connected to the E-E and IND board, to remove the camera section from the lower portion of the case.

> When performing these procedures, use care not to damage the wires and flexible cables.

After removing several small screws, (Fig. 7-12) the case can be split in two parts for servicing.

Figure 7-13 shows the locations of the important components in a VHS-C camcorder deck that might require cleaning and replacement. If the tape transport does not wind, rewind, stop, or start properly the various sensors might need to be cleaned, adjusted, or replaced.

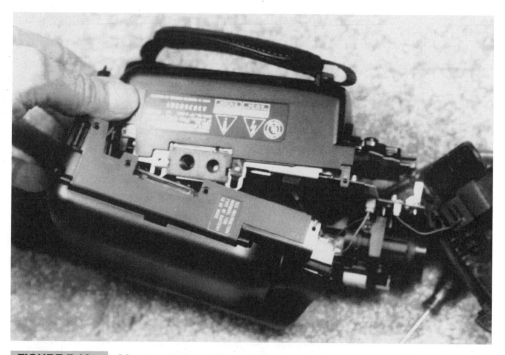

**FIGURE 7-12**    **After several small screws are removed, this camcorder case can be split apart.** Zenith

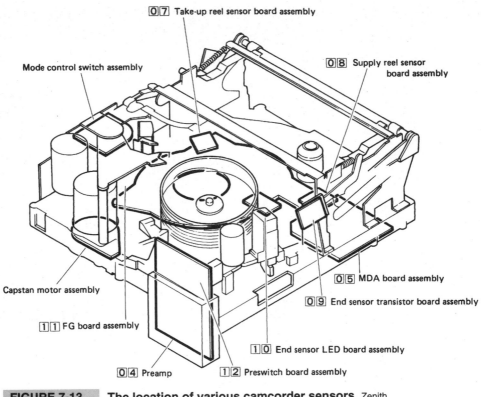

**FIGURE 7-13**   **The location of various camcorder sensors.** Zenith

**Cleaning the camcorder heads**   You should clean the camcorder heads when the picture playback becomes snowy, noisy, fuzzy, or has streaks across the monitor screen. If you keep the heads clean, it will make them last longer and save on major repair cost later. The playback picture with streaks (Fig. 7-14) was caused by a very dirty video cylinder head. Oxide build-up on the tape head cylinder could cause the tape to be pulled, thus causing damage to other mechanical parts. The heads can be cleaned with a spray cleaner, cleaning fluid on swab, or a cleaning cassette. Head-cleaning spray is used in Fig. 7-15 and a cleaning cassette is shown in Fig. 7-16. However, a cleaning cassette might not clean the heads thoroughly. You can buy cleaning cassettes that will also clean the tape guide, spindles, and the rubber rollers. It's much better to use a swab soaked in a good head-cleaning fluid. On some camcorders, you can clean the cylinder head through the open cassette lid, but you can do a better job if you remove the door cover.

On most models, you can remove the cassette cover door and then get at the machine's mechanism for repairs and cleaning. You will usually just have to remove two small screws and the cover will come off. Some units have small plugs over the screws. Notice that the door has been removed for cleaning in Fig. 7-17.

With the cover removed, you will see a large shiny drum or cylinder that rotates. The tape heads are located on this drum assembly. A swab soaked in cleaning fluid is used in Fig. 7-18. The heads can be cleaned with a spray cleaner. These cans will usually have a small tube that you can use to control the area that you need to clean. However, use

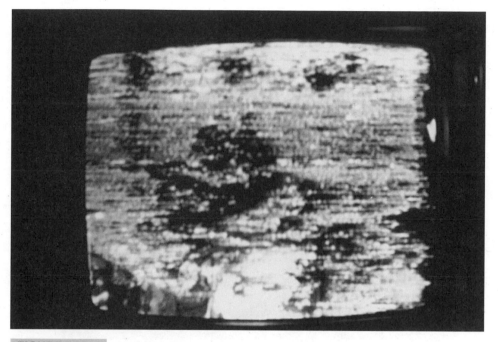

**FIGURE 7-14**    A dirty cylinder head will cause a picture that is noisy, has streaks, or a very snowy picture.

**FIGURE 7-15**    Audio/video cleaning spray is being used to clean the cylinder heads.

**FIGURE 7-16**    A video head-cleaning cassette.

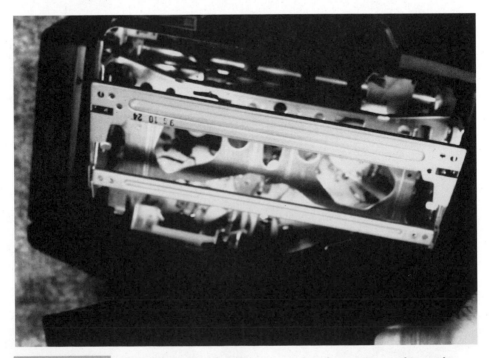

**FIGURE 7-17**    The camcorder tape door cover has been removed to make cleaning the heads and other parts easier.

**FIGURE 7-18**    A sponge swab soaked in cleaning fluid is being used to clean the cylinder heads.

caution, do not get spray into other parts of the machine. The spray can technique will not do a very good or lasting cleaning job.

If you use your camcorder a lot, you should clean the heads several times a year. As previously stated, the swab or chamois and cleaning fluid is the best technique. Do not use ordinary cotton swabs because the cotton material will pull and might damage the delicate heads. A lint-free cloth soaked with alcohol is good to use when cleaning other parts of the machine's rollers and tape guides. A cassette tape cleaner is used in Fig. 7-19, and the cleaning fluid is being applied to the cassette.

When using the swab or chamois, rotate the cylinder head from right to left several times. Always move the swab horizontally and not in a vertical motion so as not to damage the head's small tip assembly. And keep your fingers off of the drum because your body oil can cause damage. Always thoroughly clean all oxide dust from the cylinder heads and drum surfaces.

A dirty tape path, coated with oil and grease, could cause the tape to pull tight and break and/or wrap around moving parts. Some of these contaminations might even cause the camcorder to "eat" the tape. A defective cassette might have caused the tape to break, too. After the tape and cassette has been removed, thoroughly clean all of the rollers, capstan, drum, and tape path real good with an alcohol-soaked swab or a clean cloth. If the tape is so tangled up and you cannot remove the cassette, then remove the tape cover and the side or bottom covers of the camcorder. Remove any tape wrapped around the drum head and capstan. You might want to rotate the flywheel in a back-and-forth direction to remove the

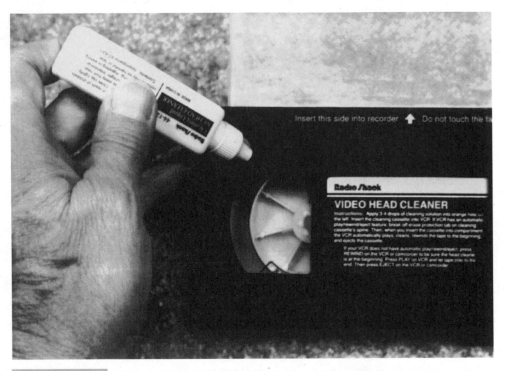

**FIGURE 7-19**    Cleaning fluid is being applied to a video head cleaner cassette.

tangled tape. You might find many turns of tape around the capstan. I have had to cut the tape with a razor blade to remove the tape. Use care in doing this! Figure 7-20 shows the oxide dust and dirt being removed from the tape guides, rollers, and head drum after the cassette was removed.

**Tape will not move and no viewfinder picture**    When you go to record and the tape will not move and all systems seem dead, you might have a defective power supply and/or a dead battery. First, check that the battery is charged and that it is installed properly. Figure 7-21 shows the battery being installed properly. Also, be sure that the battery contacts are clean. Now see if the low-battery indicator light comes on. Some camcorders might not have this indicator. If the indicator shows that the battery is low, plug in the ac adapter/charger, if you have one, and see if the camcorder now operates. If it does, check to see if the battery is bad or if the charger unit is working. A dc voltmeter can be used to see if the charger unit is supplying the correct voltage. If these items check out and there is still no operation, suspect that a fuse is blown or the power switch is dead. The motor could also be defective.

**Camera auto-focus operation**    The video camera uses infrared light rays to automatically keep the picture in focus. The infrared rays are generated on the front of the camera by an infrared LED and projected to the image that you are taping and then they are reflected back to a set of two photodiodes, also located near the camera lens. The photodiodes detect the infrared rays reflected back from the object and a time-lapse circuit

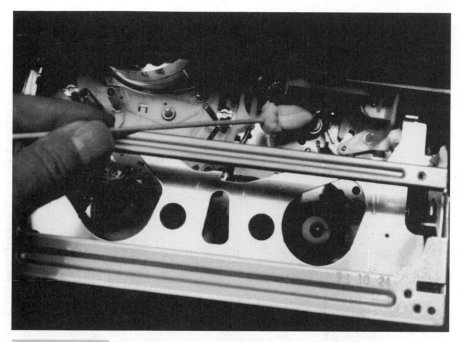

**FIGURE 7-20**     The various rubber rollers and tape path guides should also be cleaned when the tape head is cleaned. You can use isopropyl alcohol for this.

**FIGURE 7-21**     Be sure that the battery is installed properly and that the contacts are clean.

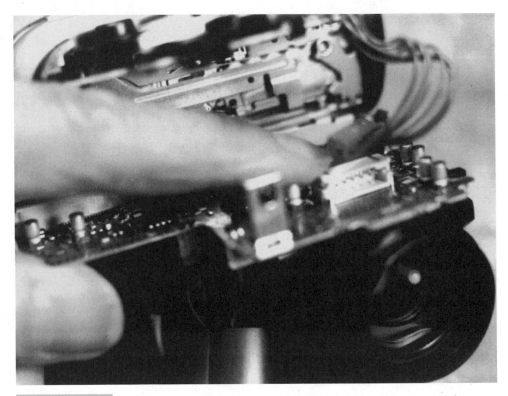

**FIGURE 7-22**   Inspect the zoom/focus circuit board for any poor connections and clean any flex cable connector contacts with isopropyl alcohol.

calculates the current needed for a control circuit to "tell" the focus motor its running time for correct focus. You can check the infrared LEDs for emission by using an infrared detector card.

The auto-focus motor obtains it control voltage from an auto-focus IC processor, usually located near the camera lens. The focus/zoom circuit is shown in Fig. 7-22. For auto-focus problems, check for loose connections or dirty cable pin connections on this board. If it seems that the auto-focus circuit is working, check for voltage to the focus motor. If this voltage is present, suspect the motor is faulty. Also, be sure that the focus sensor, located near the lens, has not been broken or damaged.

Some camcorders can be focused manually or by an auto-focus control. The lens can be rotated manually for good focus or you might want an out-of-focus picture for some special effects. When in the auto-focus mode, the lens rotates automatically. If the lens cannot be moved automatically or manually, suspect jammed gears, faulty focus motor, or even the control circuits. With the cover removed (Fig. 7-22), inspect the motor drive for stripped gears or dirt, grease, etc., jamming the gears.

**Slide switches and control buttons**   Camcorders usually have several buttons for control and zoom and sometimes slide switches for power, etc. These buttons and switches might stick or not work properly. Some are mounted under a plastic membrane and can be taken apart for cleaning. The surface-mounted switches can be cleaned with a spray switch cleaner. When cleaning, slide the switch back and forth a few times to clean the contacts and work in the spray cleaner. This cleaning fluid will also solve sticky button problems

**Cassette not loading properly** Some camcorders load the cassette tape electronically when you place the cassette in the holder or press a button. In other models, you press a button to release the holder door and then close the door manually to load the tape. If the tape will not load properly, you might have a defective cassette or some object might have gotten inside the camcorder via the loading door. Check and see if the cassette holder door has not been bent, damaged, or broken. This could happen if the camcorder was dropped.

When the cassette is inserted, you should hear the loading motor working to pull in the cassette on such models. If you do not hear this, suspect that the loading switch is bent or out of adjustment. Also, check for a loose or broken drive belt or a broken or jammed loading gear assembly. The loading gears are being inspected in Fig. 7-23.

**Intermittent or erratic operation** If your camcorder has intermittent and erratic performance in various operational modes, such if it will not load or eject the cassette, it might not record or play back tapes at times, intermittent zoom lens operation, or at times, it will not work at all, then suspect a faulty flex cable with intermittent open wire runs or poor/dirty flex cable pin contact connectors. These flat flex cables are very delicate and the contact pins (Fig. 7-24) are very small. Clean the cable contacts and flex the cable while

**FIGURE 7-23**   Inspect for a jammed or broken loading gear assembly if the tape cassette will not load into the camcorder.

**FIGURE 7-24**    If it is having intermittent camcorder operation, check the flex cables and any plug-pin connections. Clean all connections.

operating the camcorder modes and see if the problem comes and goes. If it does, you might have to replace the flex cable.

**Camcorder motors**  In most camcorders, you will find three motors: one each for the capstan, cylinder drum, and zoom. Some of the older, more-sophisticated camcorders, have motors for loading, auto focus, and iris adjustment. However, some of the small 8-mm camcorders will only contain the loading motor, capstan, and drum motors.

If your camcorder has variable speed, the problem is usually a dirty or loose drive belt. Check the belt for being stretched, dirty, cracked, or worn. You can remove the belt and clean it with alcohol and a clean cloth. If this does not help your speed problem, then replace the belt with the correct one. If a belt has been slipping, it will usually appear very shiny on the pulley side. And be sure that you clean the idlers, pulleys, and capstan thoroughly with alcohol and a swab before replacing the new belt. A good cleaning and new belts will solve most erratic camcorder speed problems. Other erratic camcorder speed problems can be caused by a defective motor, defective or misadjusted brake-release system or, on some machines, a clutch pad release.

**Camcorder troubles and solutions**  Many camcorder troubles are caused by some minor faults that you might be able to correct yourself with the following list of troubles and solutions.

*Symptom/trouble*   Will not record or playback.

*Probable cause and correction*   Check for a cassette in the unit. Try ejecting and reinserting the cassette. Cassette may be defective. Is dew indicator flashing? If it is, moisture is in the camcorder. Keep it at room temperature for a few hours and retry. Also, the cassette tape might be at the end. Rewind the tape and try again.

*Will not playback*   The VCR Play button must be in the VCR (Play) mode.

*Symptom/trouble*   Will not record. The safety tab on the cassette is closed.

*Probable cause and solution*   The VCR Play button must be in the Camera mode.

*Symptom*   No picture is in the viewfinder.

*Probable cause and solution*   The lens cap might still be over the lens.

*Symptom/trouble*   Camcorder turns itself off.

*Probable cause and solution*   Turn camcorder back on with the power switch. Some camcorders will turn themselves off if left in the Record or Play Pause mode for three minutes to prevent tape wear.

*Symptom/trouble*   Camcorder will not work with remote control unit.

*Probable cause and solution*   Be sure that the remote control is aimed at camcorder's LED sensor. The lithium battery might be dead or not installed correctly in the remote-control unit. The remote-control sensor on the camcorder could be exposed to direct sunlight or strong artificial light.

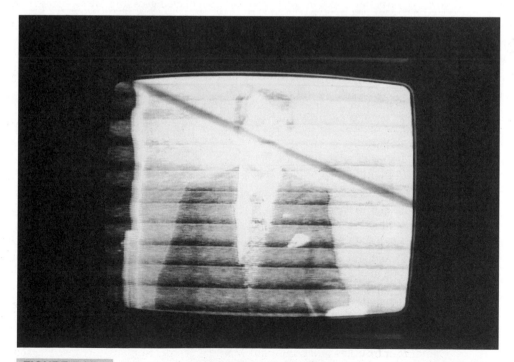

**FIGURE 7-25**     **Noise bars (streaks that are caused by a very worn tape cylinder head drum.)**

*Symptom/trouble*   Sound is very low or distorted on playback.
*Probable cause or solution*   The person you are recording might be too far from the camcorder. Some camcorders have an external microphone that can be installed for greater range of audio pick up.

*Symptom/trouble*   Very poor auto-focus operation.
*Probable cause and solution*   The object that you are taping might not be in center of the viewfinder or two objects are at different distances. If your camcorder has a focus-lock feature, turn it to on.

*Symptom/trouble*   The viewfinder displays are out of focus.
*Probable cause and solution*   Be sure that the lens is clean and not smudged. The eyepiece focus control could be misadjusted. Some camcorders have three small control adjustments near the viewfinder. These are brightness, color, and focus. These might need to be readjusted. Use caution when adjusting these because they are miniature controls.

*Symptom/trouble*   While recording, the camcorder will unload and then shut-off.
*Probable cause and solution*   The tape has come to its end. Check for a defective tape-end switch.

*Symptom/trouble*   While recording, the color recorded is different than the actual color. The color is not true.
*Probable cause and solution*   Adjust the white balance.

*Symptom/trouble*   The picture is blurred when played back.
*Probable cause and solution*   The cylinder head is dirty, worn, or defective. See Fig. 7-25 photo.

*Symptom/trouble*   The external microphone is not working.
*Probable cause and solution*   Check and clean the microphone switch. Check for a broken microphone cable or plug connection.

*Symptom/trouble*   No picture is on viewfinder screen during tape playback.
*Probable cause and solution*   Be sure that the TV/video switch is in the Video position.

*Symptom/trouble*   The tape will not fast forward or rewind.
*Probable cause and solution*   The drive belt is very loose or broken, the drive belt is slipping, or the tape is at its end.

*Symptom/trouble*   The cassette starts to load, but then immediately shuts the camcorder down.
*Probable cause and solution*   The tape has not engaged properly. Try reinserting the cassette. The cassette tape has come to its end.

*Symptom/trouble*   The ac adapter/charger has no ac operation.
*Probable cause and solution*   If you find, with an dc voltmeter, no or low voltage, check for an open fuse, faulty diodes, or a faulty transistor regulator. Check the ac line cord and plug. Check the power transformer winding for an open with an ohmmeter. Clean all switches and contacts.

*Symptom/trouble*   The ac adapter/charger will not charge the battery. The battery might be defective. Try a new one.

**FIGURE 7-26**    **A magnification lens and light are being used to locate poorly soldered joints or cracked PC boards, which can cause intermittent camcorder operations.**

*Probable cause and solution*    Check the adapter's output voltage. If it's OK, the charging circuits might be bad. If the charging LED will not light, the charging circuits are the prime suspects, also. Check the transistors, LED charge indicators, and the zener diodes in the charging section of the adapter. If the charging circuits and voltage output is good and the battery does not hold its charge very long, the battery is defective.

*Symptom/trouble*    Intermittent video recording. When you are looking into the electronic view finder (EVF), the picture intermittently goes blank, streaks, or breaks up.
*Probable cause and solution*    With the camcorder case removed, try tapping the various sections of the PC board with a pencil eraser and see if the picture breaks up on the EVF. If it does, it might mean that a poor solder connection or a crack is on the printed circuit board. With a magnification light (Fig. 7-26), you might be able to locate the defect and repair it.

# Camcorder Care Tips

- If possible, store your camcorder and tapes at room temperature.
- Always replace the lens cap when not using your camcorder.

■ Before using your camcorder, be sure that your hands and face do not have any chemical residue, such as suntan lotion, because this could damage the unit's finish.

■ Keep dust and dirt from getting inside the cassette door. Dust and grime are abrasive and will cause wear on the camcorder's head drum, gears, belts, and cassettes.

■ The camcorder can be damaged by improper storage or handling. Do not subject the camcorder to swinging, shaking, or dropping.

■ When the camcorder is not in use, always remove the cassette and ac adapter and/or battery.

■ You should keep the original carton if you need to ship it for repair or to store it.

■ Do not operate the camcorder for extended periods of time in temperatures below 40 degrees F or above 95 degrees F.

■ Do not aim or point your camcorder at the sun or other bright objects because this could damage the CCD imager.

■ Do not leave your camcorder in direct sunlight for extended periods of time. The resulting heat buildup could permanently damage the camcorder's internal components.

■ Do not operate your camcorder in extremely humid environments.

■ Do not operate your camcorder near the ocean because salt water or salt water spray can damage the internal parts of the camcorder.

■ Do not use an adapter, adapter/charger, or batteries other than those specified for the camcorder. Using the wrong accessories could damage the camcorder.

■ Do not expose the camcorder or adapter to rain or moisture. If either component becomes wet, turn off the power and dry out or have it checked out by a service company.

■ Avoid operating your camcorder immediately after moving it from a cold location to a warm location. Give your camcorder about two hours to reach a stable temperature before inserting a cassette. When the camcorder is moved from a cold to a warm area, condensation could cause the tape to stick to the cylinder head and damage the head or tape. Some camcorders have a dew indicator and the machine will not operate until the moisture has been eliminated and the temperature has stabilized.

# 8

# WIRED TELEPHONES, CORDLESS PHONES, ANSWERING MACHINES, AND CELLULAR PHONE SYSTEMS

# Telephone System Overview

The telephone (telco) company lines from your home or office go to a central office or to a telco substation (also called *switching* or *call-transfer stations*). The calls over a pair of copper wires (twisted pair) are one of many pairs within a telephone cable. This cable could be located overhead on poles or buried underground. Most telephone calls are carried over fiberoptic cables from the substations to the central switching office. Fiberoptic cables are used between cities and other points across country. In a few selected areas, fiberoptic lines are used from some home and office phones.

From your local telephone central office your calls will go to routing centers to be transmitted across country to another area to complete your long-distance calls. These calls can go over AT&T long lines via copper wires, coaxial cables, microwave signals, satellite signals, or fiberoptic cables.

## TIP AND RING CONNECTIONS

Regardless of what type of phone you use, the signal will start out over two copper wires. These two wires are called *ring and tip* on the phone plug and jack connections. Proper telephone wiring designates that the *tip* is the green wire and the *ring* is the red wire.

## THE TELEPHONE RINGER (BELL)

The telephone "ringer" device is used to let you know when an incoming call is coming onto your phone line. When you have a call, the central switching office sends out a burst signal of approximately 85 to 110 peak-to-peak ac volts at a frequency of about 20 Hz. This burst lasts about two seconds and is off about four seconds. The old ringer bell uses two electromagnetic coils of wire that move a metal clapper back and forth from the magnetic force. The clapper hits two metal gongs, thus providing the ring. This is called an *electromechanical bell ringer*.

Most modern electronic phones use an IC to produce a more pleasing electronic ring. This electronic ringing sound is produced by a piezoelectric device or tone generator.

## THE HOOK SWITCH

To clean and repair a conventional Western Electric telephone, you must remove two screws on the bottom (Fig. 8-1). Also note, in the upper left side of this photo is a thumb wheel that is turned to control the loudness of the phone bell ringer.

With the top cover removed (Fig. 8-2), you can now clean or repair most parts of this phone. The hook switch is being cleaned and burnished in Fig. 8-2. You can use very fine grit sandpaper or an emery board to clean these switch contacts. In this phone, the switch has several contacts that are pushed down by the weight of the handset that it is hung upon the cradle. The hook switch connects or disconnects the phone's voice circuit from the telephone line. Dirty or corroded switch contacts can cause noisy reception, intermittent phone operation, or complete loss of telephone operation. The old-style mechanical rotary dial on some phones also have contacts that might need to be cleaned.

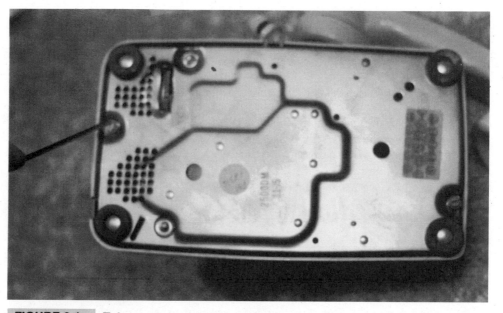

**FIGURE 8-1**    Take out two screws from the bottom of a conventional Western Electric phone to clean and repair the unit.

**FIGURE 8-2**    The "hang-up" switch contacts are being cleaned in a conventional phone.

## TELEPHONE HANDSET AND TOUCH-TONE PAD

The phone handset usually houses the transmitter and receiver units that is connected to the main phone base unit with a coiled flexible cord. Because the cord is flexed a lot, it might fail and need to be replaced. The module plugs and the connectors that they plug into might become dirty and need to be cleaned. Some phones also have the Touch-Tone pad built into the handset. The Touch-Tone pad produces a unique *dual-tone multifrequency* tone *(DTMF)* that does all of the tone signaling in place of the old rotary dial pulse system. ICs are used in conjunction with the Touch-Tone pad to produce the DTMF tones.

# Conventional Telephone Block Diagram

A conventional phone block diagram is shown in Fig. 8-3. The ringer (bell), hook switch, and dialer have been covered thus far. The dialer can be the old rotary dial or the modern touch-pad (DTMF) tone system. Now look at the speech circuit and see why it is needed. The speech circuit is used to couple the receiver and microphone transmitter in the handset, which has four wires into two wires for connection to your phone line. This is needed for full-duplex operation for you to talk and listen at the same time and use only one pair of wires. It also couples the dialer into the phone lines and produces a sidetone to the receiver that lets you control your speech level. The speech network consists of a hybrid transformer and a balancing network. A hybrid speech circuit is shown in Fig. 8-4.

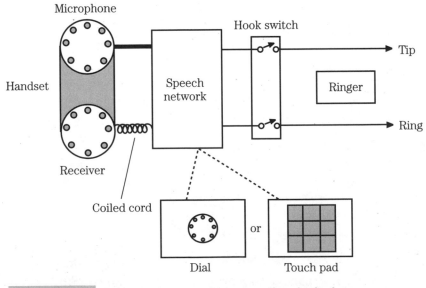

**FIGURE 8-3**  **A block diagram of a conventional telephone.**

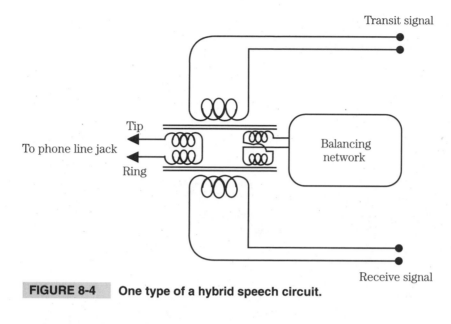

Transit signal

Tip

To phone line jack

Ring

Balancing
network

Receive signal

**FIGURE 8-4**    **One type of a hybrid speech circuit.**

# Some Conventional Telephone
# Troubles and Solutions

If your phone is not working (dead), then check that phone jack with another working telephone. If that phone is also dead, you need to go outside your house and locate the phone box
or telephone network interface housing. Figure 8-5 shows one type of telephone housing box.

## USING THE TELEPHONE TEST NETWORK BOX

The telephone network interface test and connection box is provided so that you or a telephone technician can determine if the phone problem is in the home wiring, jacks, or the
phone lines. This box has a convenient test jack that will help you to isolate telephone line
troubles. You need to make this test before reporting a trouble to your phone company.
This could save you an unnecessary dispatch and service charge.

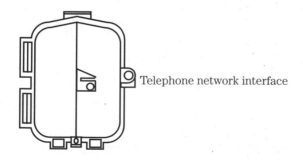

Telephone network interface

**FIGURE 8-5**    **A telephone
network interface housing this
test box to isolate your phone
problems.**

To make this test, remove the modular plug from the interface box jack and insert the plug from a working telephone set (Fig. 8-6). Now try your telephone. If a dial tone is heard, the problem is in your home phone equipment or house wiring.

After you have finished your phone test, unplug the telephone set (Fig. 8-7) and reconnect the modular plug back into the interface jack. Close the cover and screw the fastener down until the cover is snug and tight.

Now that you have confirmed that the phone line is OK, go back into your house to be sure that all telephones, answering machines, cordless phones, fax machines, DBS receivers, and computer modems have been unplugged. A problem in any one of these units could cause your complete phone system to fail. With all phone items unplugged, you can then plug an operating telephone back into a jack. If this phone now works, one or more of your other units you have disconnected is faulty. To find out which one, plug one item in at a time, then check your phone. If your test phone stops working, you will know which of your other phone equipment is faulty.

## STATIC AND PHONE NOISE CHECKS

If your phone has static and noise, plug in another phone for a test. If the test phone is noise free, then your original phone is faulty. If you still have noise on the test phone, the wall jack might need cleaning or you have wiring problems. If the phone is faulty, then switch the line cord and clean any module connections. Next, check and/or switch the handset cord and clean any plug-in connections. If you still have noise, take the phone apart and

**FIGURE 8-6    Modular plug being removed from the interface jack.**

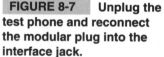 **FIGURE 8-7** **Unplug the test phone and reconnect the modular plug into the interface jack.**

clean any switch contacts and look for loose connections. If the phone has a printed circuit or flexible circuit wiring, inspect it for poorly soldered connections and cracks.

## LOW SOUND OR DISTORTION

For these symptoms, always check the cords, plugs, and any switch contacts. Another item to check is the transmitter and receiver diaphragm, located in the handset. The earpieces and mouthpieces can be removed, like you would take off a jar lid. They have threads on the caps that can be unscrewed. Clean the metal diaphragm and the electrical contacts. The microphone or receiver units might be defective and need to be replaced.

## DTMF TOUCHPAD PROBLEMS

When any type of liquid (Fig. 8-8) is spilled into a telephone or other electronic equipment, various problems can develop. If any liquid gets into your telephone touch pad, you should immediately take it apart and flush it out with water or a good electrical contact cleaner. Then use a hair dryer to dry out the touch-pad circuit boards (Fig. 8-9). You can check the operation of the Touch-Tone pad by lifting the receiver and listening as you push

**FIGURE 8-8**    Liquid spilled into a phone can cause many problems.

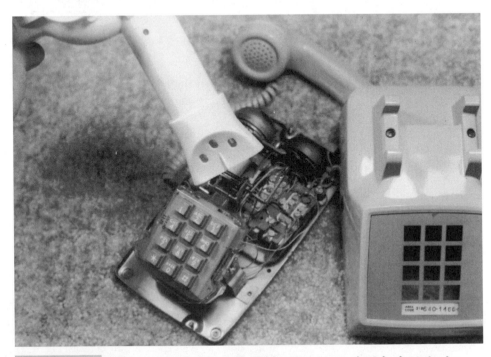

**FIGURE 8-9**    After cleaning the phone with water or an electrical contact cleaner use a hair dryer to dry out the phone circuits.

each button. You should hear a different tone for each button pushed. If you do not, clean the button contacts on the membrane pad and check all wiring from the pad unit to the main circuit board. The pad might have to be replaced or other components on the main board might be faulty if the tones are still not produced.

# Electronic Telephone Operation

The electronic telephone contains diodes, capacitors, resistors, ICs, PC boards, and many other components. Refer to Chapter 1 for more details on these discrete components. Figure 8-10 shows the block diagram of the electronic phone, including a dialer IC, speech-network IC, ringer transducer, Touch-Tone keypad, and a voltage-regulating power supply. In most cases, the dc voltage is taken from the phone line to power the phone circuits. These phones will feature volume and voice level adjustments, multiple ringing, phone number memory bank, last-number-dialed memory, and many other features. Some advanced features are hands-free speaker phone systems, LCD display readout, and Caller ID.

The speech-network IC block (Fig. 8-10) is an IC that receives and transmits speech and the DTMF tone signals. The speaker/microphone is usually a electrodynamic type.

A zener diode protection device across the phone-line input is used to protect the phone circuit from voltage spikes and surges. The ringer IC is connected directly across the phone line and has a dc block to prevent loading down the telco line.

Most electronic phones have a dual-mode IC. This mode switch is labeled $T$ and $P$. When in the T position, the IC sends out DTMF tones for dialing and in the P position, the old-type pulses are sent out for dialing.

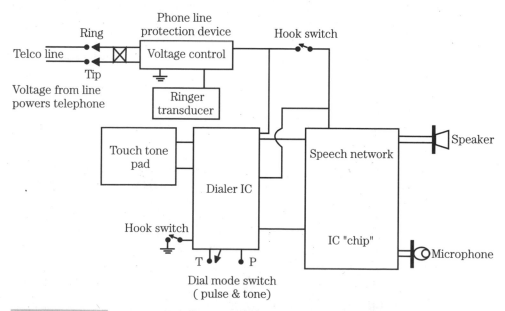

**FIGURE 8-10**    Basic block diagram of an electronic telephone.

**FIGURE 8-11**    An Western Electric Princess phone taken apart for cleaning and repairs.

Most all components in an electronic phone can be mounted on one or two PC boards. Figure 8-11 shows a Western Electric Princess phone that has been taken apart for repair and cleaning. Notice one circuit board is in the base and another one is mounted in the handset.

Some phones use a microprocessor to enhance its functions and capabilities. With a microprocessor, many more features can be added, such as a visual display for clock time, Caller ID number display, call waiting, call transfer, call restrictions, answering-machine control, and many other features. These phones are very complicated and you should have them repaired at an electronic service company that specializes in telephone repairs.

## ELECTRONIC TELEPHONE TROUBLES AND REPAIR TIPS

Electronic circuits, as well as parts layout, vary from one model of phone to another. As this section covers various troubles and solutions, you might want to refer back to Fig. 8-10.

**Noisy phone operation**    A typical electronic phone with the cover removed from its base is shown in Fig. 8-12. If you have any noise, popping sounds, or intermittent phone operation, you should check, clean, and tighten all of the screw terminals shown. Also, clean and tighten the module phone jack connection shown on the bottom in Fig. 8-13 or on the back or sides of other phones. Phones that have a Touch-Tone pad in the handset also have a pushbutton hook switch next to the pad. If these button slide switch contacts become dirty, it can cause noisy or intermittent phone operation. This switch and spring (Fig. 8-14)

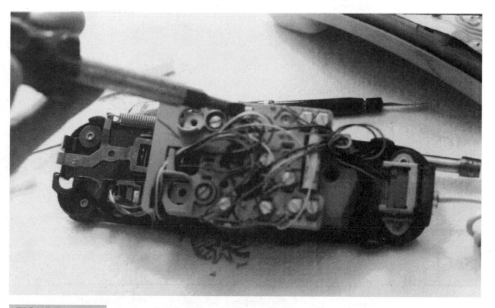

**FIGURE 8-12**    A circuit board located in the base of an electronic phone. Clean, check, and tighten any loose screw connections.

**FIGURE 8-13**    Clean the module jack with a good contact cleaner or with isopropyl alcohol.

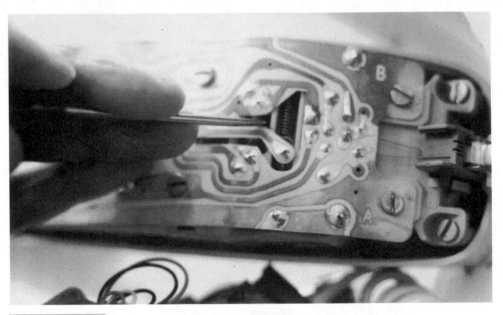

**FIGURE 8-14**    Location of the hook switch slide contacts.

is located under the PC board. Use a spray contact cleaner for cleaning these slide switch contacts.

**No phone operation (dead)**   Make these checks after you have determined that another phone works OK in this phone jack location. Some phones will have an external (power block) dc voltage supply that plugs in the ac wall outlet with a cable that then plugs into the back of your phone. Be sure that this power block in plugged in and then measure for 9 to 12 Vdc at its cable plug with an voltmeter. Also check the coiled cord that goes to the module phone jack at the phone's base and then connects to the handset. Phones with an internal built-in power supply might have a blown fuse or defective surge suppressor that will cause the phone to be dead. Make a visual inspection for loose connections or poorly soldered joints (Fig. 8-15). A small hair-line crack on the PC board is often hard to locate, but can cause all types of telephone problems. A good bright light with some magnification can help you to locate these PC board defects. A small hair-line crack on the PC board is shown in Fig. 8-16. This is the circuit board you will find when the handset cover is removed. With the handset apart, check the wiring solder connections and tighten all screw terminals, resolder or repair the connections, as needed. Clean the receiver and transmitter elements and their electrical contacts.

**Touch-Tone pad problems**   Most telephones with a touch pad have provisions for tone and pulse dialing modes. These modes are selectable by a slide switch labeled *pulse* or *tone*. Be sure that this switch is in the Tone mode if you do not hear tones as you dial. The switch might be in the center slide position and cause a no tone or a pulse dialing condition. The slide switch contacts might be dirty and need to be cleaned.

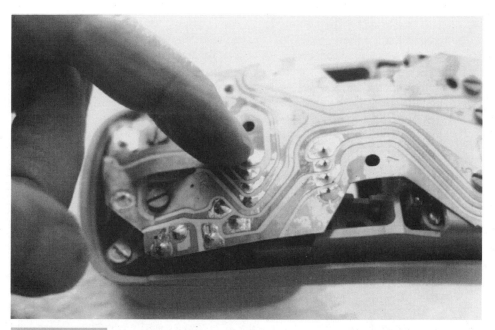

**FIGURE 8-15**    A poorly soldered connection might cause the phone to be dead or operate intermittently.

**FIGURE 8-16**    A PC board hair-line crack is hard to locate, but can cause all sorts of phone problems.

If one or more tone buttons do not work, remove the phone cover and remove the keypad so that you can check all wiring and connections to this unit. Repair any broken wires or poorly soldered connections. Also look for any broken or damaged membrane switches. The keypad will often become dirty because liquids have been spilled into the pad. Carefully take the keypad membrane assembly apart and clean and dry out all components within the pad. You might have to replace the pad assembly if it is too badly damaged, cracked, or broken.

# How a Phone Answering Machine Works

Telephone answering machines are found in many homes and offices. Of course, some people will not talk to a machine, but most find it an indispensable product. It's a time saver for receiving calls when away from home and screening those many nuisance calls when you are at home.

There are basically two types of answering machines. The old models that use one or two miniature cassette tapes. Figure 8-17 shows a cassette being installed in the tape compartment drawer. The unit can be a complete phone/machine combination (Fig. 8-18) or a

**FIGURE 8-17**    **A cassette being inserted into the tape compartment of an answering machine.**

**FIGURE 8-18** A typical stand-alone tape answering machine.

machine that you can plug into your existing phone. The other type is the tapeless answering system that uses ICs to digitize and store the phone messages into IC memory and synthesized speech circuits.

A typical answering machine that uses one miniature cassette tape is shown in Fig. 8-19.

## CONVENTIONAL TAPE MACHINE OPERATION

Your answering machine will have to detect a ring signal from the central office in order to tape a message. The ringer circuit detects and sends this incoming ring to a ring-detector circuit. This circuit converts the analog ring signal to digital logic for counting. This ring logic is counted by the microprocessor or CPU by the number of rings you select; this starts the machine tape with your prerecorded message.

# CONTROLS AND FEATURES

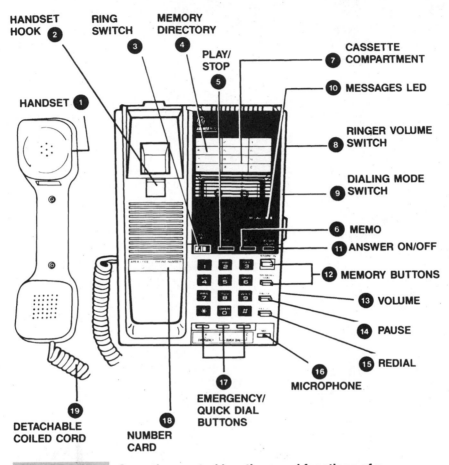

**FIGURE 8-19**   **Operating control locations and functions of a conventional answering machine that uses a cassette tape.**

When the correct rings are detected, the CPU "tells" a relay to close, which seizes the phone line, starts the tape recorder, and connects the speech network. Figure 8-20 shows a simple block diagram of a single tape answering machine. The incoming call is amplified and you can hear the person calling, which enables you to screen the calls. Also, a microphone built in the machine allows you to record the outgoing messages (OGM).

Most late-model machines have a built-in DTMF decoder so that you can control your machine from any telephone by calling your home phone number. This DTMF decoder is connected to logic circuits and controls the CPU to give your machine the desired instructions. The CPU or microprocessor is the large-scale IC (LSI), with 36 or 42 pins (Fig. 8-21). By a preset code, you can retrieve your answering machine messages when away from home. This is called a "beepless" remote-control system. With older machines, you carried a handheld beeper to control the machine over the phone system.

**Play/record operation**   Briefly, the component that actually handles the recording and re-playing of tape messages is the play/record (P/R) amplifier (Fig. 8-20). The P/R amplifier controlled by the CPU is what "tells" the amplifier if it needs to handle outgoing or in-coming messages and should the tape recorder be in the Play or Record mode. A beep tone

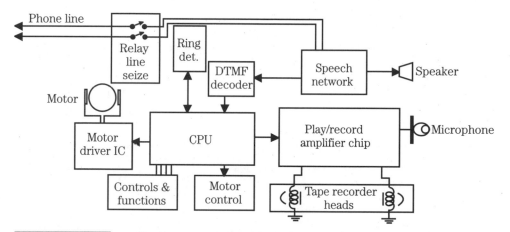

**FIGURE 8-20**    **A simplified block diagram of a "tape type" answering machine.**

**FIGURE 8-21**    **The LSI CPU or microprocessor found in an answering machine usually has 36 or 42 pins.**

controls the operating status with a logic signal to the CPU. The beep detect lets the machine find the beginning and end of each message.

**Cassette tape operation overview**   The job of a tape player in an answering machine is to run the magnetic tape back and forth at the proper speed and record and playback the correct portions of the tape as instructed by the CPU. In most tape machines, the drive motor (Fig. 8-22) is used to operate all of the mechanics for tape operations. A small rubber belt from the motor drives the cassette hub gears and the capstan. The hub gears are shown in Fig. 8-23, with the cassette tape removed.

**Cleaning the tape mechanical system**   For good record and playback operations, the tape must be kept at a constant tension and travel. To do this, a pinch roller is pressed against the capstan shaft with the tape passing between them. There can be one or more idler rollers. The pinch and idler wheels are made of rubber. The capstan and pinch roller is shown to the left of the pencil in Fig. 8-24 and the record/play head is to the right. All of

**FIGURE 8-22**    The main drive motor uses a small belt
to operate all of the tape-recorder mechanics.

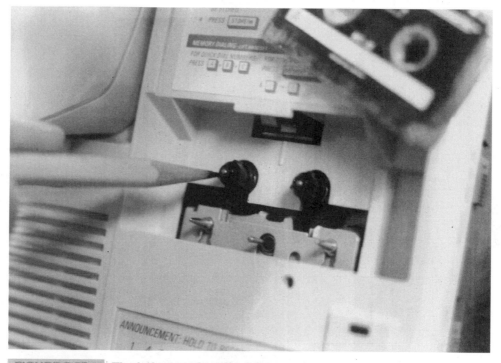

**FIGURE 8-23**    The hub gears turn the tape spools within the cassette.

**FIGURE 8-24**    The capstan and pinch roller is to the left of the pencil and the record/play head is on the right.

**FIGURE 8-25**    Use a small brush and alcohol to clean all tape recorder mechanical parts.

the rubber rollers, capstan drive shaft, record/play head, and tape guides should be cleaned with denatured alcohol or a nonsolvent cleaner. After a period of time, all of these parts will have a build up of oxide from the tape. Dust and dirt should also be cleaned out with a small brush and alcohol from all of the tape mechanism (Fig. 8-25). As pointed out in Fig. 8-26, the record/play head can be cleaned with a cotton swab and alcohol without removing the top case cover from the answering machine.

For more information and service tips on the tape recorder section, refer to Chapter 2 on audio/stereo cassette player systems.

**Digitized tapeless answering machines**    Many answering machines do not use a cassette tape, but utilize memory chips, analog-to-digital conversion (A/D), and speech digitizing processes. An answering machine that uses the digitizing system is shown in (Fig. 8-27) and is used with a conventional telephone plugged into the unit.

Refer to the simple block diagram in Fig. 8-28 to see how the tapeless machine works. These machines use a digitized speech network that records both outgoing and incoming voice messages. Some of these units can hold several outgoing messages that can be changed with a button touch, many incoming messages, a time/date stamp and you can rapidly select which message you want to hear in any sequence.

**FIGURE 8-26**    The record/play head can be cleaned without removing the top case of the answering machine.

A microphone is still used for your outgoing messages and these electrical signals are digitized as well as the incoming signal message on your phone line. These analog speech signals go to an analog-to-digital converter (ADC). The ADC samples the analog signal at a very high rate and converts it into digital words. To recover the incoming and outgoing messages, the digitized data from the memory chip must go to the digital-to-analog converter (DAC) for you to hear the messages. After filtering from the DAC, the reconstructed synthesized voice is very near that of the original. These tapeless machines can have more features and are almost trouble-free, compared to the cassette answering machines.

**Some answering machine troubles and solutions**    To repair and clean the answering machine, take out the screws (Fig. 8-29) to remove the bottom of the case.

*Problem or symptom*    Machine will not answer an incoming call.
*Probable cause and correction*    Check the ring-detection circuit. Also check and clean the cord and module plugs from phone jack to the answering machine base. Figure 8-30 shows some zener diodes that are in series with the input phone line for protection. They might be defective because of lightning surges and could lower the ring voltage level. Use an ohmmeter to check these diodes. The problem could also be that too many phones are on one line, which can reduce the ring voltage level.

**FIGURE 8-27**    **A tape-less or IC memory/digitized speech answering machine.**

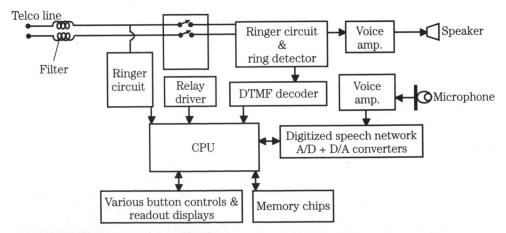

**FIGURE 8-28**    **A block diagram of an all-electronic digital answering machine.**

**FIGURE 8-29**    **For cleaning and repair, remove four screws from bottom of the machine.**

*Problem or symptom*   Cassette tape will not rewind.

*Probable cause and correction*   Check for a loose or broken belt from the motor to the hub spindle. Clean any dirt or grease from belts. Also, a broken or jammed gear could be the trouble. Check and clean any dirt or grease from the mechanical parts or rubber wheels.

*Problem or symptom*   New messages are being taped over all old messages. The message is not intelligible.

*Probable cause and correction*   The tape is not being erased between recording sessions. The erase head or its circuitry and wiring are usually at fault. Also, be sure that the erase and play/record heads are clean.

*Problem or symptom*   Tape will not move or moves erratically.

*Probable cause and correction*   You might have a broken tape or a damaged cassette. Remove cassette and replace with a new one. Check gears and spindles to see if they turn freely or are jammed. If tape is broken and tangled check the capstan shaft and pinch roller and see if any tape has been wrapped around them. Remove any tape and clean these components.

**FIGURE 8-30**    These zener diodes might be faulty and cause the ring-detection circuit not to work and the machine will not answer or record calls.

*Problem or symptom*  No dial tone.
*Probable cause and correction*  Check all phone cords and module plugs. Check hook (receiver hang-up) switch for proper movement.

*Problem or symptom*  The message indicator flashes, but no message is recorded.
*Probable cause and correction*  Replace the cassette tape with a new one and record another message. Clean the record/play heads.

*Problem or symptom*  Loss of memory modes.
*Probable cause and correction*  The small battery (Fig. 8-31) could be worn out. This battery usually plugs into a slot on the bottom of the machine and should be replaced. If the battery has become corroded, the connectors should be cleaned. Take the case off of the machine and clean it with a brush and solvent (Fig. 8-32).

**FIGURE 8-31**     The small battery used for chip memory is being replaced.

**FIGURE 8-32**     Battery terminals being cleaned with a soft brush and solvent.

*Problem or symptom*  The answering machine will not function. It beeps and the call-counter LED flashes.

*Probable cause and correction*  Machine has locked because of a loss of ac line voltage, surge, spikes, etc. Unplug the power block from the ac outlet for 20 seconds, then plug back in. This will reset or reboot the microprocessor (URT) within the answering machine.

*Problem or symptom*  The message sound level is too low.

*Probable cause and correction*  Check the setting of the volume control. Check and clean the play/record head and capstan.

# Cordless Telephone Overview

The sales of cordless phones probably account for over half of all telephone sales. These phones set you free to roam around room-to-room, all over your home and even outside in the yard and workshop, etc.

## SOME CORDLESS PHONE CONSIDERATIONS

Some portions of the cordless and conventional corded phones have the same operations. They both convert the sound of a voice into electrical signals and transmit them via telephone lines to another telephone receiving set. At the same time, the telephone converts these electrical signals of the person's voice "at the other phone" back into sound waves. Of course, the big physical difference with the cordless phone is that there is no cord between the handset and phone base.

With a conventional phone, the electrical impulses are carried by the cord between the handset and the phone base; then they are sent out over the telephone lines. However, with a cordless phone, the electrical signals travel between the handset and telephone base via radio waves.

The cord from handset to base has been replaced by a two-radio, which has duplex operation and allows two conversations simultaneously. A simple cordless phone drawing is shown in Fig. 8-33.

As with two-way radios, auto radios, and CB radios, the reception and interference can vary from location to location and from time to time. These same kinds of problems can be a factor with many cordless phones. This could be bothersome because we have all expected very clear reception over the fine telephone systems. Americans now expect phone privacy, excellent sound quality and high reliability.

### Some cordless phone problems

*Poor sound quality*  Some phones might have poor audio response, receive interference from electrical devices and interference from other cordless phones close by.

*Short range*  The main appeal of a cordless phone is the ability to let you move around without pulling a cord. However, some phones have a very short range.

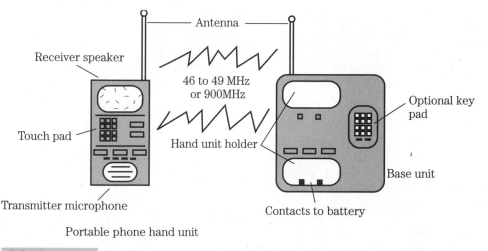

Antenna

Receiver speaker

46 to 49 MHz
or 900MHz

Optional key
pad

Touch pad

Hand unit holder

Base unit

Transmitter microphone

Contacts to battery

Portable phone hand unit

**FIGURE 8-33** Basic cordless phone operational.

*Your phone use time is limited* Because the cordless phone is powered by a battery in the handset, the phone might quit during a conversation because the battery needs to be charged. Also, most cordless phones will not work if the home power goes off.

*Conversation privacy* Other people with a cordless phone on your frequency can listen to your conversation if they are within range of your phone. However, the newer cordless phones, some in the 900-MHz band, offer digital transmission with encoded speech information and also automatic channel switching if someone transmits on your frequency.

**Cordless phone frequency bands** The early model and some even sold today work on the 46- to 49-MHz radio-frequency band. This is a small region between the CB band and TV Channel 2.

A new generation of cordless phones were developed in 1990 in the 900-MHz band. These phones operate at a much higher frequency (902 MHz to 928 MHz) and a greater transmitter output power. Phones operating at these higher frequencies have less interference and the band is not as crowded. Lucent Technologies (formerly AT&T) and Panasonic now have 900-MHz phones with a range of up to 4000 feet.

**Two transmission modes** Not only do cordless phones operate in two frequency bands, but they have two different transmission modes to transmit and receive conversations. Early model phones used analog transmission, which is a continuous signal that varies in intensity like a radio broadcast station. An early model analog phone is shown in Fig. 8-34 photo. The latest 900-MHz phone technology uses digital transmission, which is a series of short, computer-coded signals that are decoded at the phone's receiver. Digital phone transmission reduces the noisy, buzzing, and crackling usually found in cordless analog phones; also, they are harder for someone to eavesdrop on.

**FIGURE 8-34** This cordless phone uses analog RF signals with FM modulation on frequencies from 46 MHz to 50 MHz.

***Digital phone modes*** The digital phone manufacturers use different ways to transmit their encoded narrow-band signal in the 900-MHz frequency range. Spread-spectrum cordless phones stretch (or spread) the narrow band signal over a multitude of different frequencies and are not as susceptible to interference. Spread-spectrum phones would be like many people talking identical messages all at the same time over the phone system. If one, or even quite a few, of these messages are blocked or interfered with, you would still receive the message from the others. Because the signals transmitted from these cordless phones are "spread out" over a wide bandwidth and with increased transmit power, these phones will have increased range and voice clarity. The 900-MHz phone (Fig. 8-35) has a range of more than one-half mile.

## SOME DIFFERENT PHONE TECHNOLOGIES

Now look at some of the various cordless phone technologies that deal with clarity, privacy, and range.

*Call security* The new digital technology now makes it possible to eliminate the eavesdropping problem.

**Some security codes now being used** To keep outsiders from using your cordless phone, almost all phones sold today use some type of code between the handset and the phone base to prevent unauthorized use of your phone line by someone using another handset on your frequency. However, many phones don't secure the call itself.

*Basic scrambling*    This is a basic way of scrambling a conversation so that it is more difficult to decipher, except by its own receiving set. Just about any competent electronics technician could unscramble this code.

*Digital encoding*    This one generates coded signals that are more difficult to decipher than a basic scrambling mode. Probably a competent electronics engineer could unscramble this coding and reconstruct the conversation.

*Spread-spectrum encoding*    This is the most difficult code to crack, having been developed by the military for war applications. It would probably take a few years for a highly skilled communications engineer, with very specialized knowledge and very sophisticated equipment, to eventually bypass this security code. This spread-spectrum encoding goes by a trade name of *Surelink Technology.*

**FIGURE 8-35**    A digital 900-MHz cordless phone that has a one-half mile range or more.

## CORDLESS PHONE SOUND QUALITY

The characteristics of these two phone frequency bands has a lot to do with the quality of reception. Cordless telephones in the 46- to 49-MHz range are much more prone to interference from a much more cluttered frequency band. This band is much more susceptible to electrical interference and other radio services. Cordless phone receivers in the 900-MHz band must be a lot closer to an interference source to be affected.

*Companding*  This system is somewhat like the Dolby stereo audio system. The system essentially "loudens" the transmission to overcome the naturally occurring hiss, then brings it down to normal levels at the receiving end. This technique goes by trade names of *Sound Charger* and *Compander*.

*Multichannel capability*  Many cordless phones can operate over several channels. When you hear some interference, you can manually switch to another channel. However, some automatically move to another channel, looking for a channel with less interference. Because of the limited space between channels in the 46- to 49-MHz range, this technology is not very effective for these cordless phones. In the present FCC frequency band allotment, the number of channels is limited to 25 for low-band phones, 40 for high-band standard, and digital cordless phones have the equivalent of 100 high-band spread-spectrum channels.

*Digital transmission*  This transmission involves sending the message as a series of computer codes. Because each bit of code only has a designated value of "1" or "0," unlike analog radio transmissions that have infinite possibilities—the receiving set can more easily identify the incoming code in the presence of interference. However, if there is significant interference, the conversation might sound "choppy" because an entire code is lost.

*Spread-spectrum transmission*  The radio transmission technique also uses a series of computer codes. However, because the same signal is stretched out over a broad frequency band, the likelihood of "choppy" conversations is considerably eliminated. A receiver only needs to receive a part of the transmitted signal to reconstruct the original message. Spread-spectrum transmission will retain its quality—even if the 900-MHz frequency becomes more crowded. Digital spread-spectrum cordless phones often display the Surelink technology label.

*Cordless telephone range*  Range is a key characteristic of the cordless telephone—regardless if it operates in the low or high radio band and whether it is analog or digital. Different cordless phones transmit various levels of power, much the way radio stations use different levels of power output. Higher-powered cordless phones can transmit signals over greater distances. However, the phone also needs more battery power to do this. This will require larger batteries or a shorter use time. Spread-spectrum cordless phones get additional range at lower power levels because they use battery power more efficiently than nonspread-spectrum phones. The FCC also allows the spread-spectrum phones to operate at higher transmit power levels than conventional phones.

*Analog phones*  These phone systems have the shortest range and are the most likely to be affected by high buildings, hills, etc. These analog cordless phones, operating in the crowded low band, are restricted by the FCC to no more than 0.04 milliwatts of radiated power, and rarely exceed 500 feet of working range.

*Standard digital phones*  The inherent characteristics of this mode of transmission, plus the fact that they typically transmit in the 900-MHz band, increases their range up to 0.25 mile. The FCC limits the power output of these phones to no more than 0.75 mW.

*Spread-spectrum digital* The spread-spectrum system improves on the advantages of standard digital transmissions because of multiple signal transmissions. The FCC also allows far greater power output (up to 1 W), which have ranges of up to 0.5 mile.

*Phone battery life* The handset of a cordless phone has a battery pack for operating power, and it has to be recharged after being used for a period of time. The battery is recharged by placing the handset back on the base unit. The "talk time" for cordless phones is usually hours and the "standby" time, not being used or recharged, is in days. The time required to recharge a fully discharged cordless phone battery is approximately 8 to 12 hours. Of course, the phone cannot be used during this period. The base of the cordless phone is powered by ac power, with a plug-in power block. When you lose ac power, the cordless phone does not operate. However, if you use a *UPS (uninterruptable power supply)* you can plug your cordless phone power block in the UPS power supply and not have a loss of your cordless phone operation.

*Quick charge capabilities* Phones with a quick-charge system use more-expensive circuitry built into the phone. Most cordless phones with quick-charging features will recharge in one hour, but cost more than the standard, slower-charging systems.

*Back-up battery units* A phone-charging system for a backup battery, located in the base unit, provides a few advantages. First, the base of the cordless phone will have power if the ac power line fails. Second, the battery in the base of the phone and the battery in the handset can be "interchanged;" it will be kept charged, thus providing a "hot spare" battery for extended phone conversations.

**Deluxe cordless phone features** Now, with many cordless phones being sold, many manufacturers have opted to combine cordless phones with other telephone features, such as answering machines, speaker phones, or caller ID packages. These features can be useful, but they do not improve the sound quality of the basic cordless phone. You might want to consider buying a separate telephone with all of the deluxe features, and then buy a "stand-alone" cordless phone. One reason for this is that the more gadgets you build into one system, the more probability of a failure with a higher repair or replacement cost.

## CORDLESS PHONE BUYING TIP

Your choice of cordless phone should be decided by what you need the most: Range, security, clarity, cost, or a combination of all four. Generally, the more limited the cordless phone capabilities, the less it will cost. If you reside in uncrowded rural areas and only require a short range, you might not want to consider digital spread spectrum. However, you will receive the best performance from the cordless telephone with the most sophisticated technology. For more information on Surelink spread-spectrum phone technology, call (800) 858-0663.

## BASIC CORDLESS PHONE OPERATION

The cordless phone consists of two units that must work together somewhat like two two-way radio systems. The base unit transmits and receives RF signals and the portable battery-operated handset unit also receives and transmits RF signals.

The base unit has electronic circuits that connect into your local phone line. It also has a radio transmitter and receiver circuits. The base unit plugs into an ac outlet, usually a block power unit, to power the radio transmitter/receiver and has a built-in battery charger for charging the handset battery.

## BASE UNIT CIRCUITRY

The cordless phone not only has to connect to the phone line, but also has to have a complete radio (RF) transmitter and receiver in the base (Fig. 8-36). The base also has a CPU, memory ICs, phone-line seize relay, ring-detector circuit, ringer, and some models will have a DTMF pad, tone-generator chip, and a back-up battery for power outages.

The base unit contains five blocks. These would be the power-supply/charger circuits, speech or interface network, microprocessor (CPU) controller, the radio receiver and transmitter sections. The RF carrier with modulation, which is transmitted back and forth between the base and handset is also modulated with speech (voice) and control signals. These control and speech signals are modulated in the transmitter and demodulated in the receiver. A duplex circuit is used so that the transmitter and receiver can use the same antenna. Interference and feedback is eliminated because different frequencies are used for transmitting and receiving. The receiver also has filters in its RF stages.

## THE PORTABLE HANDSET UNIT

Figure 8-37 shows a typical cordless phone handset unit with all of the function control call-outs. As you refer to Fig. 8-38, you will notice that the handset contains most of the circuits found in the base unit. The handset has a transmitter, receiver, CPU control chip, ringer circuits, and DTMF circuits.

The CPU in the base and handset units controls all of the cordless phone operations. The CPU along with memory chips (ROMs or RAMs) keep track of all memory and program instructions for phone operations.

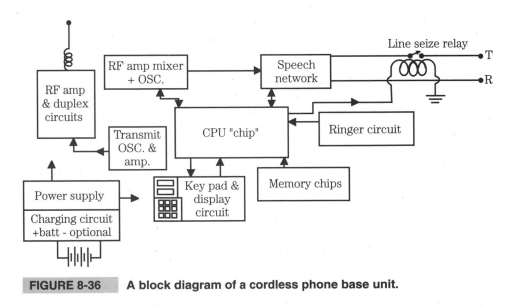

**FIGURE 8-36**    A block diagram of a cordless phone base unit.

① **Antenna**

② **Volume Control**

③ **Channel Scan**

④ **Charge Contacts**

⑤ **Indicators**

RINGER – Lights when ringer is on and phone is in use, flashes when battery is low. (Indicator flashes even if ringer is off and battery is low.)

PHONE – Lights when phone is in use, flashes for hold and mute.

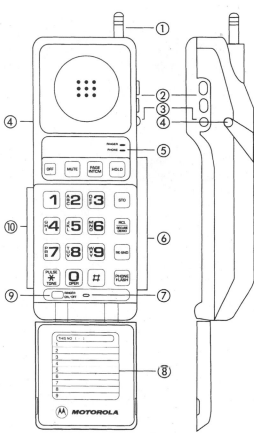

⑥ **Function Keys**

OFF    Hangs up phone when flip is open.

MUTE   Mutes so other party cannot hear you.

PAGE INTCM   Pages base.

HOLD   Places calls on hold.

STO    Stores phone numbers.

RCL SECURE DEMO   Recalls numbers from memory. Press and hold for Secure Clear™ demonstration.

RE-SND   Redials last number dialed.

PHONE FLASH   Accesses call waiting (if available). Press for dial tone if flip is open.

PULSE ✳ TONE   With phone off, press and hold to switch to pulse dialing. Press and hold again to return to tone dialing (see page 9). Changes temporarily from pulse to tone dialing during a call (see page 18).

⑦ **Microphone**

⑧ **Flip with Telephone Number Card**

⑨ **Ringer Button** – Turns handset ringer on and off. If off, you will not hear your handset ring or beep when paged. Turning ringer off extends battery life up to 21 days.

⑩ **Numeric Keypad** – Enters dialing information.

**FIGURE 8-37**    A cordless phone hand set with all of the call-out functions.

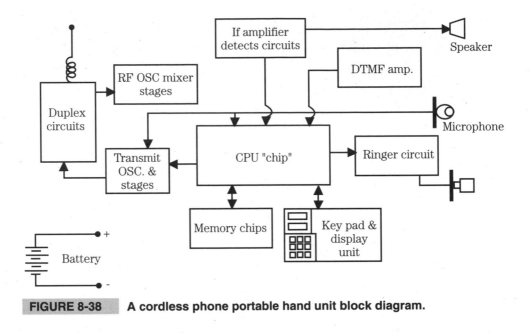

**FIGURE 8-38**    **A cordless phone portable hand unit block diagram.**

In the base unit, the CPU gives the instructions for the transmitter, the receiver, and sends control pulses to the portable hand unit, as well as interpreting control pulses sent back from the hand receiver/transmitter unit. The CPU also controls the phone-line seize relay, ring circuit detector, and DTMF dialing signals.

## CORDLESS PHONE TROUBLES AND CORRECTION HINTS

Let's now look at some cordless phone problems. Some of these problems might be caused by electrical interference, other cordless phones, two-way radio interference, weak batteries, or no battery charging, not enough talk range, or other people listening in on your phone conversations. Other problems could be your phone not working at all (dead) or an intermittent operation problem.

**Removing the phone case**    To clean or repair the phone base unit, remove four or more screws (Fig. 8-39). This will give you access to the circuit board, power supply and charger section, and phone-line seize relay.

To take the portable hand unit apart remove the battery cover and take out the battery. Then remove the two screws (Fig. 8-40). Now lift up the battery end of the cover and swing the two covers apart (Fig. 8-41). This will then expose the circuit board, switches, and other components.

Now check the solder connections on the flat ribbon cable that connects between the two covers sections. If the phone has static noise or intermittent reception, check for circuit board cracks or poorly soldered connections (Fig. 8-42). Also clean any dust or dirt from any of the small switches (Fig. 8-43) and the membrane under the touch pad.

**FIGURE 8-39**    To repair or clean the phone, remove the four screws from the base unit.

**FIGURE 8-40**    Take off the battery cover and remove the battery. Then take out the screws to separate the two handset covers.

**FIGURE 8-41** After the screws are removed, the covers will come apart.

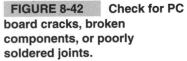

**FIGURE 8-42** Check for PC board cracks, broken components, or poorly soldered joints.

**FIGURE 8-43**    Clean dirt from the switch contacts. Use a brush or a spray switch contact cleaner.

## CORDLESS PHONE TROUBLE CHECKLIST

If you have phone problems, the following checklist should be helpful.

- Be sure that the power cord or power block is plugged in and the outlet has ac power.
- Be sure that the telephone line cord is plugged firmly into the base and the wall phone jack.
- Be sure that the base unit's antenna is fully extended.
- If the phone does not beep when the (phone) button is pressed, it could indicate that a battery needs to be recharged. Some phones have a (LO BATT) indicator light. Also, clean the contacts with a pencil eraser (Fig. 8-44) if the battery will not stay charged.
- Be sure that the battery pack is installed correctly.

**Handset and base unit not communicating (two beeps)**    Usually a "two-beep" signal indicates that the handset and base are not communicating properly. This could be caused by something as simple as being out of range when dialing a call. Try moving closer to the base unit and try the call again.

If moving closer to the base does not work, then it might be that the handset and base have different security codes. Try the following procedure:

Place the handset in the base, and check to be sure that the charging light is on. Wait 15 (or more) seconds, then pick up the handset and press the Phone button. The Phone lights on the handset and base should now go on, and the phone should now work normally.

**FIGURE 8-44** Clean these contacts if the battery will not charge
or if the phone does not work, but gives you a beeping tone.

### Phone will not work (dead)

- Verify that the modular jack is working by testing with a known-working phone.
- Is the power cord or power block plugged in? When the base is plugged in and the handset is in the base, the charging light goes on to indicate that power is connected. If the charging light does not come on, you might need to clean the charging contacts with switch contact cleaner and a soft cloth.
- Check the cordless phone line cord connected between the base and the modular wall jack outlet.
- Are both antennas pulled all of the way out? Do this for the base unit and the portable handset.
- Place the handset in the base unit to set the security code between the base and the portable handset.
- Be sure that your phone is set correctly to either pulse or tone dialing to match your local phone service.

**Noise or static problems** You are hearing noise or static when using your cordless phone. This is probably local electrical interference. Try pressing the channel button to change to a different channel.

Some phones have automatic channel-change circuits. If you still are receiving static or noise, try the following tips:

- Move the handset unit closer to the base.
- Be sure that both antennas are pulled completely out.
- Try moving the base unit to another electrical outlet. Choose one that is not on the same circuit as other appliances.

**Phone will not ring**  If the handset unit will not ring, try the following suggestions:

- Be sure that the ringer button or switch is in the On position. Some cordless phones have a battery-saver switch; when it is on, the unit will not ring.
- Check the cord and be sure that it is connected properly and that the power cord is plugged in.
- Be sure that the antenna is pulled all the way out.
- Move the handset closer or relocate the base.
- Change the channels.
- Unplug one of your other telephones. The strength of the ring signal is reduced if you have several phones on one line.

**Phone will not work (dead)**

- Unplug and replug the power cord and telephone line plug. Then pick up the handset and place it back in the base holder. If the phone still does not work try some of the following tips: Place the handset in the base and be sure that the charging light comes on. Unplug the ac adapter from the wall ac outlet, wait 15 seconds, then plug it back in again. The charging light should go on again. Wait another 15 seconds, then pick up the handset and press the Phone button. The Phone lights on the handset and base should go on, and your phone should now operate properly. If not, then go to the next step.
- Pick up the handset, open the battery compartment door and unplug the battery pack. The battery pack and small plug is shown in Fig. 8-45. Wait 15 seconds and then rein-

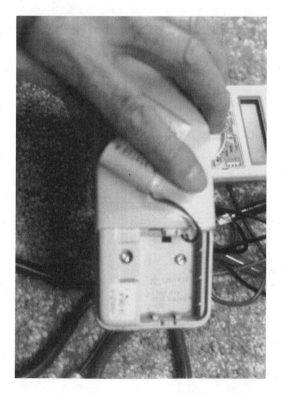

**FIGURE 8-45**    If the phone will not work, unplug the battery (small red and black wires). Wait 15 seconds and plug the battery back in. The phone should now operate.

stall the battery pack plug. Now, close the battery compartment door, place the handset in the base and check to be sure that the Charging light is on. Wait another 15 seconds, then pick up the handset and press the Phone button. The Phone lights on the handset and base should now go on, and your phone should now operate properly.

**No dial tone**   Recheck all of the previous suggestions. If you still do not hear a dial tone, disconnect the cordless phone and try a known-good phone in its place. If no dial tone is in the test phone, the problem is in your house wiring or with the local phone service. You can also plug your test phone into the outside phone junction box. If you do not receive a dial tone at this location, contact the local phone company repair department.

### Phone interference review

- Be sure that the base and portable handset antenna is not broken and is fully extended.
- You might be out of range of the base.
- Press and release the Channel Change button to switch channels. This will not interrupt your call.
- Household appliances plugged into the same circuit as the base unit can sometimes cause interference. Try moving the appliance or base to another outlet.
- The layout of your home or office might be limiting the operation range of your portable phone. Try moving the base to another location. An upper story base location will increase range.

**Cordless phone antenna replacement**   You can easily replace a bent or broken antenna on your cordless phone. Most antennas are just screwed on or off (Fig. 8-46). Most electronics stores will probably have a replacement for your model phone. Also, if you cannot get your cordless phone operating, Radio Shack has a repair service for most all brands of cordless phones.

**Phone surge protection**   A phone surge- and spike-protection module that plugs into an ac outlet is shown in Fig. 8-47. These units can protect any type phone or answering machine from lightning spikes and surges. You might want to install one on each of your phones.

# Mobile Radio Telephone Communications

For over 50 years, it has been possible to have 2-way communications via radio from a moving vehicle. This was first accomplished by a two-way radio system, then by radio telephone, and for the past decade or so with high-tech cellular devices. These cell phones can now be used just about anywhere in this country at an affordable price. And it is a great emergency device to have when you are traveling. You can think of the cell phone as a very sophisticated cordless phone system that was just explained previously in this chapter. The portable cell phone could be in your auto or coat pocket and the base unit would be the cell radio transmitter site that has a telephone interconnect to the local phone

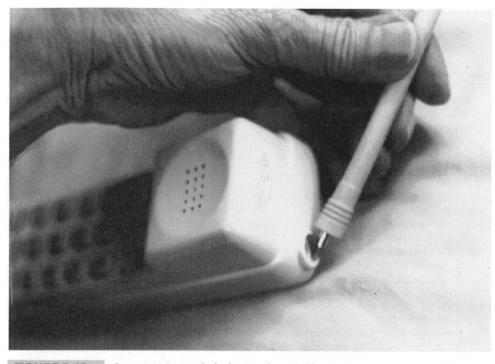

FIGURE 8-46    A new antenna is being replaced. Most phone antennas will screw in.

FIGURE 8-47    A telephone line-protection device can be easily plugged into an ac wall outlet. This provides good protection from spikes and lightning damage coming into the phone line.

company land lines. These cell sites are spaced so that you usually have contact within a site's coverage area. As you drive, your cell phone signal is automatically passed along or "handed off" to the next cell site transmitter with the strongest signal strength. The cellular phone system operates in the 900-MHz frequency band.

## TWO-WAY RADIO TRUNKING SYSTEM

About 1980, Motorola introduced the 800-MHz two-way radio bunking system, which also had provisions for using the telephone in a mobile vehicle or portable unit. This radio trunking system could be used as a two-way radio to talk to a base station or as mobile units to mobile units and also had telephone interconnect capabilities. These trunking systems utilized the 800-MHz to 890-MHz allocations (Fig. 8-48). This system was the forerunner of the now very popular nationwide cell phone networks.

**800-MHz trunking system overview** The 800-MHz trunking system consists of at least four voice transmitter/receive channels, one data control channel that sends and receives data at all times and a central system controller. A simple basic block diagram of a trunking system is shown in Fig. 8-49.

As with cell phones, the trunking system mobile and portable units have individual IDs that must be programmed into the trunking system's central controller before these units can use the radio system. Thus, to bring up the trunking system, the transmitting mobile's ID must be known by the central controller and data logger. The trunked central controller is shown in Fig. 8-50. All mobile transmitted trunked calls are initiated through the central controller channel.

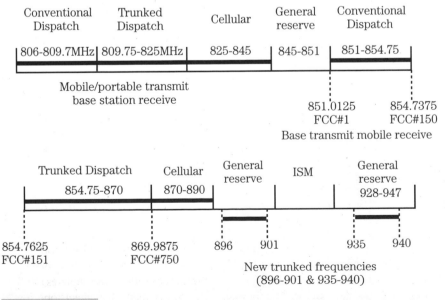

 **FIGURE 8-48** The radio trunking and cellular frequency band allocations.

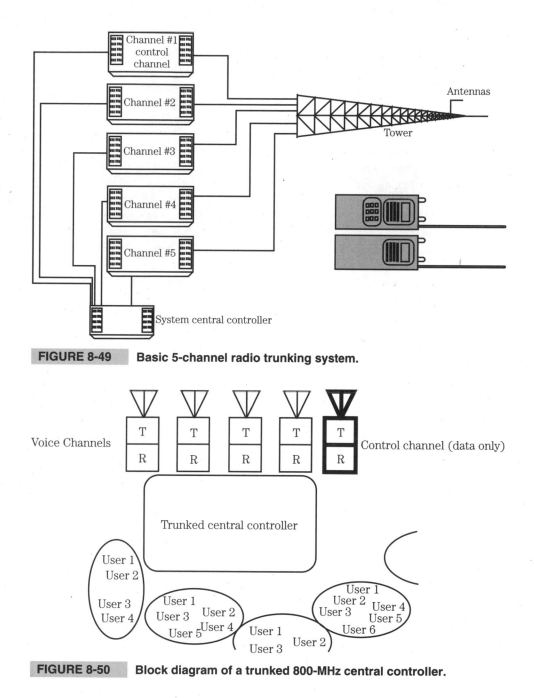

**FIGURE 8-49**    Basic 5-channel radio trunking system.

**FIGURE 8-50**    Block diagram of a trunked 800-MHz central controller.

When the mobile units microphone is keyed up and voice transmission starts, the mobile unit sends a subaudible connect tone at a low level, which the central controller recognizes and keeps the channel open or connected. When the mobile unit completes a call, a disconnect tone is transmitted and the central controller will issue a disconnect tone for the outgoing voice channel being used.

The 800-MHz trunking systems offer many features, such as private conversations (PCI, Privacy Plus) that lets a supervisor talk privately with an individual, System-wide call, Fleet-call, Status-message call, and call alert for selective paging of a specific person. These trunking systems also have a back-up called *failsafe* if the controller and data channel fail. When this occurs, the system reverts back to standard community radio repeater operation. All modes of operation are verified by data handshake signals.

**Trunking telephone interconnect**    The 800-MHz radio trunking system is also equipped so that a mobile telephone device can be used for making phone calls while on the move. These trunking systems have direct telephone interconnect with landline telephone networks. The Central Interconnect Terminal (CIT) extends the communications ability of a 800-MHz trunked radio system without changing the use of the system. Mobile or portable two-way radios that are equipped to generate Touch-Tone compatible signals (DTMF signals), can automatically access the local telephone landline system without going through a dispatcher. Also, regular phone customers who have been given an access code can call into the radio trunking system from any local or long-distance telephone system.

**The cellular telephone radio system**    The cellular radio telephone mobile operation is a highly sophisticated/complicated electronics system that has evolved over many years of development and huge investments of many communications companies. Now look at a basic cell phone system's unique operations.

The cell phone concept has very little resemblance to a conventional two-way radio communications or repeater system. Figure 8-51 illustrates how a regional cell network is laid out. The cell radio phone system breaks up the coverage area into small (cell site) divisions. Each cell site contains several low-power radio transmitters and receivers that are linked to a central cell computer controller equipment center location. When you start to use your cell phone, you will automatically be communicating directly to the closest cell site. As you travel along, your cell phone is "handed off" automatically from site to site. All of the phone calls from the cells in this group are then fed into the central cell computer-controlled switcher, which are now routed via telco lines (fiberoptics) to the local telephone exchange.

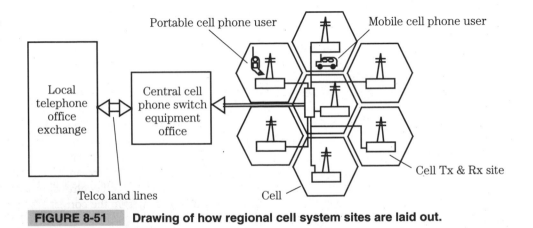

**FIGURE 8-51**    **Drawing of how regional cell system sites are laid out.**

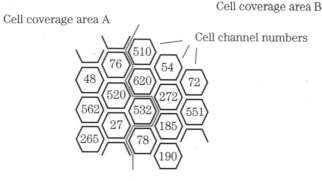

Cell coverage area A

Cell coverage area B

Cell channel numbers

**FIGURE 8-52**   How cell phone channel numbers are allocated and laid out for a typical system.

The FCC has set aside more than 600 frequencies for cellular telephone operations in the 900-MHz band. A cell site can usually handle 40 or more full-duplex telephone calls from mobile cell phones simultaneously. Each of these calls need two different frequency channels for full-duplex (two-way) conversations. Figure 8-52 illustrates how area A and area B are divided up in cells of different channel numbers and frequencies to avoid radio interference. When a cell phone customer initiates a call, the closest cell site then automatically opens up two unused channels to complete the call.

## HOW THE CELL PHONE OPERATES

The cellular phone packs a lot of sophisticated electronic components into a very small space. In fact, it is actually three devices in one unit. Not only is it a two-way radio, but a computer and telephone. And some cell devices have an answering machine and a pager. To repair them, you need to be a highly trained electronics technician and have very specialized and expensive test instruments. However, later in this chapter, some repair tips are listed that you can check on before calling on a professional service center.

**Transmit/receive section**   In Fig. 8-53, you will notice the radio RF transmitter and radio receiver sections that couple these sections with a duplexer to the antenna. The signal (voice/data) is received from the cell site and is filtered and processed to be heard in the speaker. The frequency synthesizer with instructions from the CPU tunes the cell phone to the proper receive and transmit channels. Of course, there is a touch pad and DTMF generator will enable you to make calls. Also, a read-out display indicates phone numbers that you dial and recall from memory.

**CPU and memory logic**   The heart of a cellular phone is the CPU (control/logic) and cell-control chip. The CPU receives program instructions for the ROM chip. The RAM chip is used for temporary data that is erased and updated in every day use. This could be phone numbers on a memory list, numbers to re-dial, etc. Every cell phone has an identification number and the EEPROM chip retains this and other permanent data. Not only does the cell phone process voice and DTMF tones, but it must receive and transmit a ream of data back and forth to the cell site. It also sends data to and from the cell control chip within the cell phone. The cell controller, after processing this data, sets up the correct transmit and receiver frequencies that the cell phone must operate on.

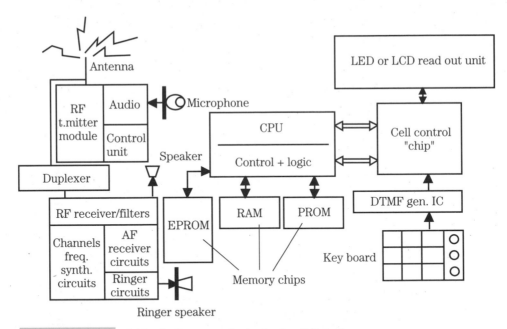

**FIGURE 8-53** **A block diagram of a typical cellular phone.**

**FIGURE 8-54** **A young lady in process of making a cell phone call.**

This should now give you a brief overview of what happens when you pick up and dial a cell phone or receive a cell phone call as the young lady is shown doing in the Fig. 8-54 photo.

**Some cell phone tips for poor, noisy or intermittent reception**  If you're using a portable cell phone, the problem could be a loose or broken antenna or your location. First, move to another location. With a mobile car phone, it could be poor coax cable, cable connections, or a broken or loose antenna. If the antenna has an on-glass mount, it might be defective or it could have been installed wrong. Do not overlook the possibility that the battery is weak or the battery contacts are dirty. Also, some computer chips in your car might cause interference to your cell phone—even if the engine is not running and the key is turned off.

**Battery talk**  Most cellular portable phones use nickel-cadmium (NiCAD) battery packs. These packs are expensive and you need to take proper care of them. These battery packs are dated by the manufacturer, and will usually be replaced if found defective and failed at an early age. Keep the phone and battery in a cool place, if possible, and keep them properly charged. Read the instructions that come with the battery pack. Most are designed for a quick-charger system. The newer NiCADs do not have the memory problem like the older batteries. However, it is best if you can discharge them completely before recharging. You might want to keep a spare charged battery with you. These batteries do wear out after many charge and discharge cycles. Also, keep the pack dry and check/clean the contacts on a regular basis.

**Drop-out and dead reception areas**  The ultra-high frequency of the trunking and cellular radio systems (800 MHz to 1000 MHz) are close to a "line-of-sight" RF signal transmission. For this reason, drop outs (loss of signals) occur when you are around hills, bridges, and large buildings. Your signal might fade in and out or flutter. In the worst case, your call might get disconnected. After you use your phone a while, you will get to know the various poor-reception areas. The dead zones will usually last longer when traveling in mountains, through hills and valleys, and in the larger cities with skyscrapers. These dead spots also depend on the proximity of the cell sites.

# 9

# COMPUTER OPERATION, PROTECTION, AND MAINTENANCE

# Computer Power Source and Protection

We must all tolerate that the power companies cannot always supply the clean, consistent power needed by today's high-tech electronics. Thus, the electronic equipment owner must take some precautions.

One study by IBM noted that a typical computer could be subjected to more than 120 power problems every month. These power problems can be minor, such as keyboard lockups, or catastrophic, with complete data loss or burned PC boards and damaged ICs. These power problems can cause a lot of downtime.

Unfortunately, business and home personal computer users are depending more on a utility power supply that is being pushed beyond its capacity. Even with the advances in modern PCs, a momentary power outage is still all that's required to lose your data. More devastating is the loss of previously written files, or even a complete hard disk, which can occur if a power problem occurs while your computer is saving a file. Network file servers, constantly writing to disk, are very susceptible at this time.

It has been noted that the two types of computer users are those who have lost data because of a power problem and those who are going to.

## POWER PROBLEMS

The following is a list of various power problems that could occur:

- *Sags*  These line voltage sags are also known as brownouts. These are short term decreases in ac line voltage levels. This is the most common type of power problem.
- *Blackouts*  This is a total loss of utility power voltage. These could be caused by too much power demand on the grid system; lightning, wind, or ice storms; other power-line damage; and power company rationing plans.
- *Spike*  The spike is also called a *voltage impulse*. A spike is an instantaneous, dramatic increase in voltage. A spike can enter electronic equipment and damage, or completely destroy, components.
- *Surge*  This is a short-term increase in voltage, typically lasting at least $\frac{1}{120}$ of a second. A surge can be caused by high-current devices being turned on and off the power-line system.
- *Noise*  The technical terms for noise are *electromagnetic interference (EMI)* and *radio-frequency interference (RFI)*. This electrical noise disrupts the smooth sine wave that you would expect to find from the utility power line service.

## UNINTERRUPTIBLE POWER SOURCE (UPS)

Because you have invested thousands of dollars in your computer, printer, and fax machines, it would be very wise to spend another hundred dollars or so for an *uninterruptible power source (UPS)*. Not only will it save your written files, but save your computer from some very expensive damage from lightning strikes or various power-line glitches.

A typical UPS, manufactured by American Power Conversion, is shown in Fig. 9-1. These units can be purchased at computer stores or Radio Shack, to name a few places. I highly recommend that you power-up your computer from one.

Figure 9-2 is a power strip that has built-in surge protection for ac power, but also the phone line that plugs into your computer modem for on-line network service.

Before we delve into computer operation and problems in this chapter, look at the UPS power system and see how one could save you a lot of computer problems before they can occur.

**FIGURE 9-1**     A UPS power-protection unit for a PC.

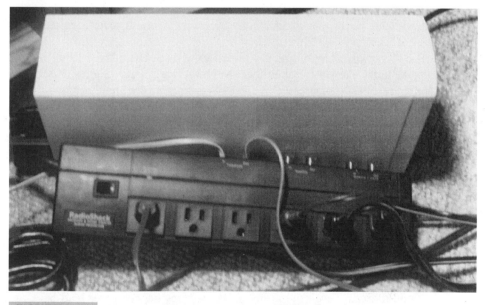

**FIGURE 9-2**     A power strip connected to a UPS unit.

**Uninterruptible power systems—fact or fiction?**  If you want to get very technical, no device can supply uninterruptible ac power to your computer. In any power supply system, the power fed to any load depends on the type of load. A computer is a special type of load that draws power in pulses that occur every 120 times per second. In reality, the computer power supply contains switches that are called rectifiers (diodes), which disconnect the computer from the ac power line socket for about 70% of the time during computer operation. Thus, it is not true that computers need continuous power (current), but the current is interrupted 120 times every second. So, in reality, there is no uninterruptible power source for computers because computers only draw interrupted power. You could then say there is a continual interrupted voltage supply from the UPS unit.

All computers have a capacitor that is used as a rechargeable battery. It operates the computer (including the disk drives) during the times (120 times per second) when the input power is interrupted. The capacitor has a very limited capacity and cannot run the computer for more than about 50 milliseconds without being charged back up. The good news is that the ac power line restores the charge of the capacitor with a pulse of power every 8.3 ms (120 times per second), about six times more than is usually required.

**How the UPS power supply works**  Figure 9-3 shows the main components that make up your UPS unit. Just follow along with the block numbers as the UPS inner-workings are explained.

***Noise and surge suppression block***  You will usually find, at the ac input of the UPS unit, a high-performance EMI/RFI noise and surge-suppression circuitry to protect the computer system. The neat thing about a UPS device is that it continuously provides suppression—even when it is turned off. The UPS suppresses noise and line surges and you really do not know it because the UPS does not transfer the computer load to its internal power source. What it does do is suppress and reduce the amplitude of noise and surges to a level much below that which can be tolerated by very delicate computer and electronic devices.

The scope waveform in Fig. 9-4 illustrates what a typical "med-level" amplitude surge or spike looks like when riding in on the power line voltage. Surges up to 15 times larger than this are easily suppressed by the UPS. Surges are commonly caused by nearby lightning strikes and heavy motor load switching on and off, such as air conditioners and heat pumps.

Figure 9-5 shows a scope waveform view of EMI/RFI noise riding in on the power line voltage. The UPS "filters" out this noise with components whose electrical resistance is very high at these radio frequencies. EMI/RFI noise is commonly created by the same activity, which causes surges, but can also be caused by nearby radio/TV transmitters, medical equipment, blinking neon bulbs, and advertisement signs.

***Ac line-voltage transfer switch***  The ac line voltage change-over switch (block 2) in Fig. 9-3, is an electric relay that quickly changes over your computer equipment from the ac power line to the UPS's internal power source whenever the utility power goes off. The UPS unit will make a beeping sound as long as the power is off. When the ac power comes back on to a safe level, the switch relay switches the computer load back to the utility power line. Except for the unit's control switches, the relay transfer switch is the only

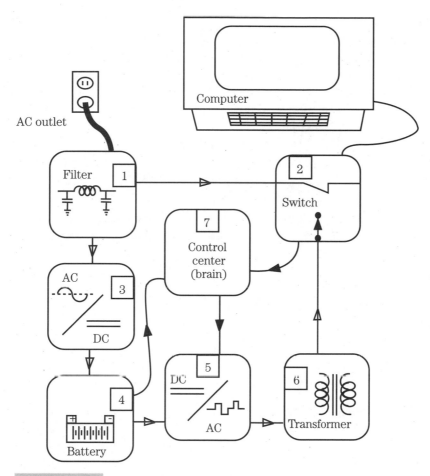

**FIGURE 9-3**    A block diagram of a UPS protection device.

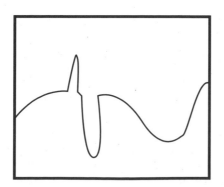

**FIGURE 9-4**    A typical "medium" amplitude spike or surge on an ac power line (as would be viewed on an oscilloscope).

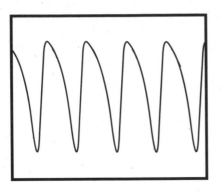

**FIGURE 9-5** What an EMI/RFI noise signal looks like on voltage from a utility line (as shown on an scope).

moving part in the UPS. The time required for the relay to switch over to either power source is considerable faster than is required for any computer or other electronic device you might have connected to the UPS.

**Power converter** The UPS has to convert its internal battery energy into a form that your computer equipment can depend on during an ac power line failure. This is what the UPS's inverter performs. The built-in UPS converts the battery's dc voltage to ac by means of solid-state devices (transistors) with a process technically called *pulse-width modulation*. This technique, which powers many other electronic devices, is very efficient and thus uses very little battery power in this conversion process. Your computer can now be expected to operate for a long time before the internal battery is discharged.

**Transformer stage** The UPS's transformer, block number 6 in Fig. 9-3, "steps up" the output voltage of the inverter to the normal ac power line voltage (120 Vac/240 Vac). Also, this transformer helps to isolate your equipment from power-line surges, etc.

**Electronics control center** Block 7 is the computer-control "brain" of the UPS. This monitoring and control circuitry detects ac power-line problems, such as sags, blackouts, spikes, and brownouts. It also synchronizes the inverter's output frequency and phase to that of the power company; detects low battery voltage levels; "tells" the load transfer switch when to act; and oversees all operator controls, indicators, and computer interface functions. Now see how the UPS behaves during power failures.

**Operation during a power outage** Always on the "look out" for a power failure, such as a blackout, sag, or brownout, the UPS constantly monitors the power line voltage and keeps the inverter always ready for a "synchronous" or smooth transfer. This means that the inverter's phase and frequency (60 Hz) is adjusted to exactly match the phase and frequency of the ac power line. If the power line voltage falls outside acceptable limits, the UPS quickly transfers your computer to power derived from the UPS's battery via the inverter and transformer that was previously described. This power transfer will occur in about three milliseconds. When the UPS is in this mode, a beep will sound every five seconds that your computer is now on limited stand-by power. If the utility power does not come back on, the UPS will emit a loud tone to warn you that you have less than two minutes remaining before the UPS shuts down and will no longer power your equipment. This

Low Battery mode indicates that the UPS's internal battery is now almost discharged. The UPS will automatically shut down if the UPS is not turned off during the low battery alarm.

Any time that the UPS detects a return of normal power line voltage when using its alternate battery source, the inverter voltage will be smoothly re-synchronized to match the phase and frequency of the power line. Once synchronized, the load-transfer switch will re-transfer your computer to power supplied by the power company. After an extended utility outage, the battery charger will then recharge the battery with energy at a pace that is consistent with maximizing the service life of the battery. If the battery were charged more rapidly, the battery life would be shortened. These batteries will usually last four to six years and can then be replaced. These batteries can be found at battery or electronic stores.

**Power company problems**    One power line condition is called *sags*, which are temporary voltage reductions of the normal 120- to 230-Vac line voltage. These voltage sags can be caused by local, high current demand from air conditioners, power tools, and electric heaters. You could also have a power reduction during a heat wave to cope with huge power demands for home and commercial air conditioners.

A power line blackout is caused by a complete loss of power. Blackouts can be caused by accidents or storms. They can also be caused by overloaded "branch" circuits (wiring within a building that is fuse protected), tripped circuit breakers or poor extension cord connections. Always check the circuit breakers in your home or office.

**Other UPS uses**    You can also connect other low-current devices to your UPS power unit. Just use a multi-outlet power strip (Fig. 9-6). You can plug in your cube power supplies for

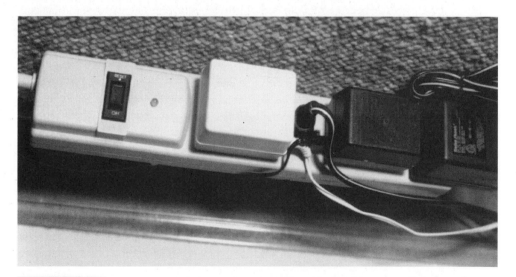

**FIGURE 9-6**    Power blocks plugged into a power strip. The strip is then plugged into a UPS unit. These blocks are used to power answering machines, cordless phones, etc.

the answering machine, cordless phone, and VCR. Then, when the power fails, you will not be disconnected from your phone call or lose any calls on your answering machine. And as a plus, it will protect these devices from lightning damage and power line surges.

### What to do if the UPS will not operate properly

*Problem*   UPS will not come on (power switch is not lit), but it beeps and the power switch (I/O) is on.

*Probable cause and what to do*

1 Line cord is not plugged in the wall socket.
2 UPS input connector is loose (230-Vac version).
3 Circuit breaker on back of unit is tripped off (reset).
4 Wall socket has no ac power. Check the building circuit breakers.

*Problem*   UPS operates normally, but the site wiring fault indicator is lit.
*Probable cause and what to do*

1 There is a building wiring error, such as a poor ground or a polarity reversal. Have a qualified electrician check and correct the building wiring.
2 An adapter plug or "cheater" might have been used and the ground connection is missing. Plug the UPS into a two-pole, three-wire grounded outlet only.

*Problem*   UPS occasionally beeps, but the computer does operate normally.
*Probable cause and what to do*   The UPS is transferring the computer to its back-up power source because of a momentary voltage sag, etc. This is normal. This audible alarm can be turned off on some UPS models.

*Problem*   The UPS beeps every so often, usually more than once or twice an hour. Your computer works OK.
*Probable cause and what to do*   The line voltage is being distorted or branch circuits are very heavily loaded. Some types of dimmer control (SCR) devices nearby can cause the same problem. Have your line voltage checked by an electrician. Operating the UPS from an outlet that is wired to a different branch circuit might help. Adjust the UPS voltage option, if it has one.

*Problem*   The UPS is emitting a loud tone. The power switch is on, but the computer is not being powered. The UPS's circuit breaker has tripped out. The ac power line voltage is present.
*Probable cause and what to do*   The UPS has shut down because of a severe overload. Turn off the UPS and unplug excessive loads. Laser printers will overload the UPS and should be plugged into a good surge suppressor. Once the overload is removed, reset the circuit breaker.

*Problem*   The UPS emits a loud tone during a power failure. The off/on power switch is on, but the computer is not powered. The UPS circuit breaker is not tripped off.
*Probable cause and what to do*   The UPS has shut down because of an overload. Turn

off the UPS and unplug any devices. Recheck the computer power requirements. The UPS can now be turned on when the power has been restored.

*Problem*   The UPS does not provide a long-enough running time. The low battery warning is noted prematurely.
*Probable cause and what to do*

1 Too much equipment is connected to the UPS. Unplug the loads from the UPS. Recheck the computer system power requirements.
2 The battery is weak (bad cell, etc.) because of normal usage or recent operation during a power failure and it has not had time to recharge. Recharge the battery by leaving the UPS plugged in for 12 hours. Do not use test control during recharge. If the UPS produces a low-battery warning too early when retested, the battery should be replaced.

*Problem*   The UPS beeps constantly. The UPS off/on switch is lit. The power line voltage is OK.
*Probable cause and what to do*

1 The line cord plug is loose. Check and/or tighten the line cord power plug.
2 The input connector for the 220-Vac version UPS is loose. Check and/or tighten the input power connector.
3 The circuit breaker is tripped. Unplug any load on the UPS and reset the circuit breaker.

*Problem*   The low-battery warning tone interval is shorter than two or five minutes, depending on the model of your UPS unit.
*Probable cause and what to do*

1 Excessive loads (current drain) are connected to the UPS output plug receptacles. The excessive loading could shorten the run time to less than two or five minutes of the low-battery warning interval. Remove any excessive loads.
2 The battery capacity is low because of frequent loss of power. The frequent power failures might not have allowed time for the battery to completely recharge, thus causing a shortened UPS run time.

*Problem*   The low-battery warning interval is much longer than the two or five minutes of normal time. This time could vary, depending on your UPS model.
*Probable cause and what to do*   Your UPS power unit is loaded to less than 10% of its rated capacity. This operation is normal. The low-battery warning interval is adjusted at the factory for consistent operation at loads above 10% of rated capacity.

# Microprocessor Operation

Before we see how a PC works, take a quick look at microprocessor operation, which is the heart of a PC. For a simple explanation of a microprocessor, use the blocks in Fig. 9-7, which shows a person who has a device that is required for a task of some sort. If the task

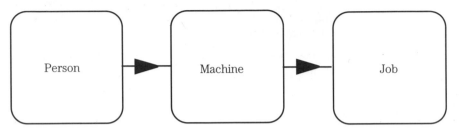

**FIGURE 9-7**    A block diagram of a simple microprocessor (CPU) operation.

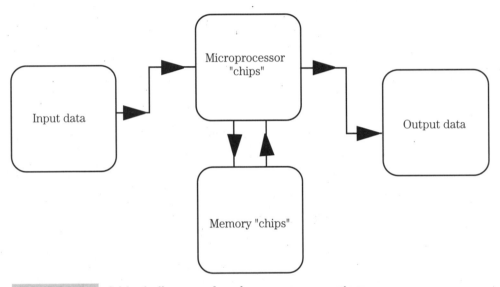

**FIGURE 9-8**    A block diagram of a microprocessor system.

is simple, such as turning on a light, the person must control the device (flip on a switch).

Note that the control information is all in one direction: from left to right. Also, if the task contains many steps or procedures, the person must remember the entire sequence to perform the operation completely.

Figure 9-8 shows the process of information flow in a microprocessor system. Again, the microprocessor receives a command from an input device (keyboard), which might be controlled by a person. However, for sequenced operations, the memory requirement is taken away from the person and transferred to the memory, which is part of the system. The input command might be only a signal, a pulse (caused by a start button), or the output could be a complete series of actions, whose codes are stored in memory. These codes are called *programs*. Determining or setting up these codes is a person, called a *programmer*.

Now find out what the microprocessor IC really does. It can perform any task that the input device or memory chips "tells" it to do. To make this easy to understand, all the microprocessor activities can be divided into two categories.

1 It dispatches data output from input or memory to output or memory. This is just a movement of data (0s and 1s) from one place to another (address), somewhat like a large multipole, multiposition switch.

2 It performs operations on data, operations, such as algebraic addition and subtraction, comparison of two numbers, and even some simple logical decision making, such as branching in a flow chart. A microchip has two sections of circuitry, a control section that takes care of housekeeping details of data routing, and an arithmetic-logic-unit (ALU) to perform the various operations.

## COMPUTER BUS ROUTES

To feed information (data) to and from the microprocessor CPU, a system of address, control, and data busses are required. These busses are a group of conductors over which bits of data are transferred to and from various points in the computer system. Some are *bidirectional busses*, which means that information (data) can be sent in both directions on these routes. In less-sophisticated designs, only one transfer of data can occur at one time for each bus. Some of the faster-model computer systems use a timesharing or bit-slicing technique that enables more than one data transfer at a time. In addition to microprocessors, many logic-designed devices use time-sharing on a single lead wire.

## HOW BUS LINES OPERATE

A digital bus is a path over which digital information is moved from any of several places to any of several destinations. Only one transfer (move) of information (data) can occur at any one time. While such a transfer is taking place, all other sources that are tied to the bus must be disabled. *Bussing* means to interconnect several digital devices, which either receive or transmit digital information, by a common set of conducting paths, over which all information between such devices is transferred.

The basic purpose of a bus is to minimize the number of interconnections (wires or PC board runs) required to transfer information between digital devices. Busses are present within ICs, microprocessors, between ICs (such as with the address, control, and bidirectional data busses present in all CPUs), and between digital systems and instruments.

Figure 9-9 illustrates a simple data bus set up for a microcomputer. The concept of a bus is probably one of the most important designs in digital electronics. In this system, all data moves via the microprocessor unit (MPU). Note that data can move in either direction between the random-access memory (RAM) chip and the MPU. All other data moves in one direction only. Data can be moved from the ROM (Read Only Memory) chip or input buffer to the MPU. And data can be moved from the MPU via latches to the outside world.

To fetch and retrieve data transfers properly (in this case, only one at a time) an address bus must be added to the system. Each data source is then assigned a different address. As an example, the RAM, ROM, input buffers, and output latches all have chip-enable pins. The correct logic pulse at these pins will activate or enable the circuit. With a different address for each circuit, then only one circuit will function at any given point in time.

In a microprocessor system, inputs to the address decoders come from the MPU via the address bus. These outputs then go to the chip-enable lines of these various circuits. Only

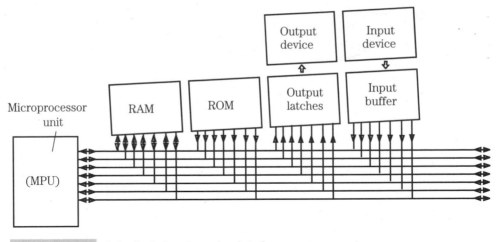

**FIGURE 9-9**    A typical drawing of a data bus arrangement.

one address can appear on the address bus at any one time, thereby enabling only one external circuit at a time.

Just as each home has its own mailing address, so must each byte have its assigned address. When any one of these addresses appears on the address bus, the RAM is selected via its chip-enable line. Notice that part of the address bus connects directly to the RAM, which selects the individual byte within the RAM chip.

ROM is also assigned a range of addresses that must be enabled when any of these addresses appear on the address bus. The output latch and input buffers are also assigned specific addresses. In this way, the MPU can "talk to" any of the external circuits just by putting the correct address on the address bus.

**Read-only memory (ROMs)**   For your PC to run its programs, the microprocessor must have read-only memory devices to give it instructions. Usually, dedicated MPUs use only ROMs. The ROM is then a permanent memory and is nonvolatile. Its contents (memory) will thus be unaltered because of power loss or system operation error. In most cases, ROMs have a random-access feature. The MPU cannot change the contents in the ROM, but can only read what has been stored.

**Random-access memory (RAMs)**   The RAM (random-access memory) chip is a device into which the MPU can write data and from which the MPU can read data while it is in the system. This is also referred to as *read-write memory*.

The RAM is a semiconductor memory into which a logic "0" or logic "1" state can be written (stored) and then read out again or retrieved. *Random* means that any one of the memory locations can be accessed by applying the proper logic states to the memory select inputs. Thus, we do not have to sequence through the memory to access a memory location. Stated very simply, RAM differs from the ROM in that the microprocessor can read from and write into it, storing data while the program is operating. Most microprocessor systems will have both RAM and ROM devices.

## HOW A PC WORKS

Now that you have the basics of how a microprocessor works, you'll now see how this all comes together to make your PC operate. Photos and drawings of the inside and outside of a popular tower PC system are shown. The section then covers PC problems that could crop up and how to solve them, also, how to read your PC monitor screen to locate problems. Included are floppy disk drives, hard drives, some tips on virus problems, and some other tips that you can use to keep your PC running in top condition.

**PC operation**    Now that you know how the CPU and busses work, see how this all dovetails together to make a PC work. As various sections of the PC is explained, follow along with the block section drawing and call outs (Fig. 9-10). These drawings and photos are of my PC (Fig. 9-11).

**Computer power turn on**    If you turn your PC on and nothing happens (it has no lights or cooling fan hum), then suspect that no ac power is getting to the PC, which could be caused by a power cord or on/off switch failure or power supply trouble within the PC. As with any electronic device, the power supply is the "life-blood" for the equipment.

Be sure the ac wall receptacle has the proper voltage. You can easily check this with a good lamp. Check for a loose power cord plug and connections and UPS trouble if you have one. Check for any blown fuses with an ohmmeter. These fuses might be inside of the PC case.

Always unplug the ac power cord before removing the case or cover. Once inside, you can check the fuses and off/on switch with the ohmmeter.

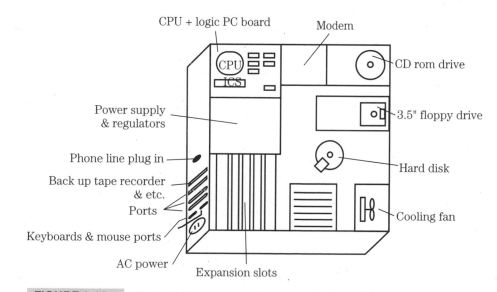

FIGURE 9-10    The inside component layout of a typical tower computer. A tower PC system gives you more room for expansion.

**FIGURE 9-11    A photo of my tower computer set-up with monitor and printer.**

Most computers use +5 and −5 Vdc to operate the CPU and the IC's functions. These voltages can be checked with a volt meter. The power supply can usually be repaired at or exchanged at a computer service center. However, you could have a fault in the computer circuits, such as a short that could have damaged the power supply. The Intel IC is shown on the motherboard in Fig. 9-12. The cooling fins for the power-supply transistors and regulators are shown in Fig. 9-13.

Also, be sure that the cooling fan is working. If it is not, your PC could overheat and cause big-time circuit damage. If the cooling fan is noisy, it could have a bad bearing or a lead wire might have gotten too close to the fan blades. If this is the case, just reposition the wires. Also, if the fan has a filter, be sure that it is clean. The fan motor is shown in Fig. 9-14.

Figure 9-15 shows the location of the plug-in strips or expansion slots that you want to use. These slots are used to plug in other boards that might be required to add other future features to your PC.

Now that the PC has power, turn it on and see if all systems are go.

**Boot-up the computer**   When you turn on the computer power switch, you are actually booting-up the computer system and letting the PC read a complex start-up and system check routine. When the computer is first turned on, the screen might be blank for a few seconds, then information will appear on the screen about the program being loaded and the capabilities of your computer. With Windows 95 (Fig. 9-16), you can stop the screen read-out by pressing F7 and check the computer options and any PC or program problems. PC booting is required so that the computer can check out all of its sections and operate

**FIGURE 9-12**    **The Intel chip located on the main PC board.**

**FIGURE 9-13**    **Cooling fins for power and regulator transistors.**

**FIGURE 9-14**    The cooling fan motor in the PC.

**FIGURE 9-15**    The expansion slots within the PC.

smoothly. This is somewhat like NASA making a count-down and checkout before a rocket launch. When loading up a Windows 95 (Microsoft) program in your PC, you will see the Microsoft logo show on the monitor screen from time to time. This is a normal occurrence. And before the PC can even load the program after turn-on, it has to be sure that

Main  Advanced  Exit

|                        |                    |               |
|------------------------|--------------------|---------------|
| System Date            | Nov 12 1996        | Help          |
| System Time            | 11:32:08           | Back          |
|                        |                    | Select        |
| Floppy Options         | Press Enter        |               |
|                        |                    | Previous Item |
| Primary IDE Master     | Conner Peripherals | Next Item     |
| Primary IDE Slave      | Not Installed      | Select Menu   |
| Secondary IDE Master   | MATSHITA CR-581    |               |
| Secondary IDE Slave    | Not Installed      | Setup Defaults |
|                        |                    | Previous Values |
| Language               | English  (US)      | Save & Exit   |
| Boot Options           | Press Enter        |               |
|                        |                    |               |
| Video Mode             | EGA / VGA          |               |
| Mouse                  | Installed          |               |
|                        |                    |               |
| Base Memory            | 640 KB             |               |
| Extended Memory        | 15360 KB           |               |

**FIGURE 9-16**    During PC "boot-up," you can stop the screen readout by pressing "F7" and check for problems, etc.

**FIGURE 9-17**    As the PC program is loaded, the Microsoft logo will come up on the screen at various intervals.

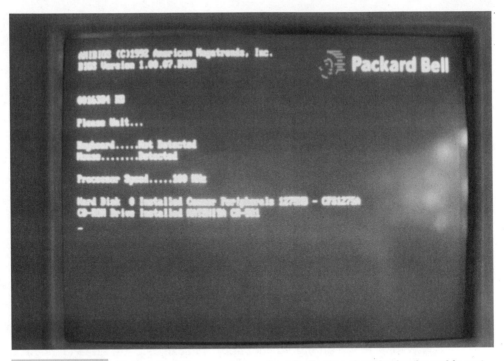

**FIGURE 9-18**   Notice that the screen readout indicates that the keyboard is not being detected.

all of the components, such as the CPU, are working and connected or plugged in. This is sometimes referred to as the *power-up, self-test.*

As the power-up, self-test is being performed, you will hear single beeps and a screen read-out that the PC is passing these tests. A lot of short beeps or long beeps probably indicates a problem has occurred and this should be indicated on the monitor screen. On some systems, no beeps are an indication of a problem. Figure 9-18 shows a screen read-out, indicating that the keyboard is not detected. Check the cable and plug-in connections.

**Cable and plug connection problems**   In some cases, a loose or dirty pin-plug connection might be the reason why your computer won't boot-up properly. Figure 9-19 shows where and how the mouse and keyboard connections can be checked. Be sure that the connections are tight. Plugging the connections in and out a few times will clean the pins. You can also use a spray cleaner solution or a pencil eraser.

The printer-port cable connections plug is shown in Fig. 9-20. Clean and/or tighten this connection if you encounter no printer action or intermittent printouts. Also, flex the cable while the printer is operating and see if the intermittent operation stops. This would indicate a defective cable.

The slots pointed out in Fig. 9-21 are used for air circulation to keep the electronic components cool. Solid-state devices do not work well when they become too hot; heat can also damage or destroy the component. Keep these slots clean and do not cover them up. These slots can also be used for circuit board additions to update or add options.

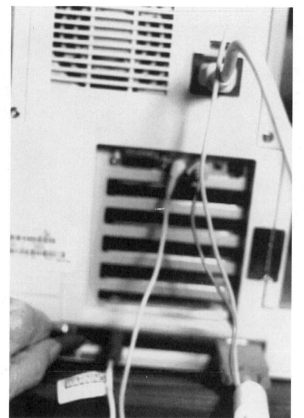

**FIGURE 9-19**    Check the keyboard and mouse plug-in cable connections for loose or dirty contacts and pins.

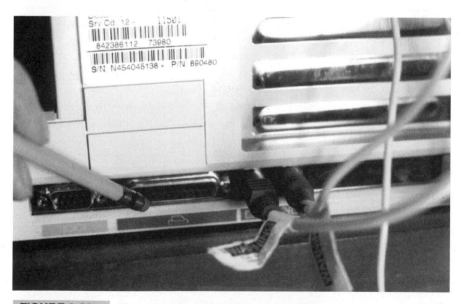

**FIGURE 9-20**    The printer port location at the PC's rear.

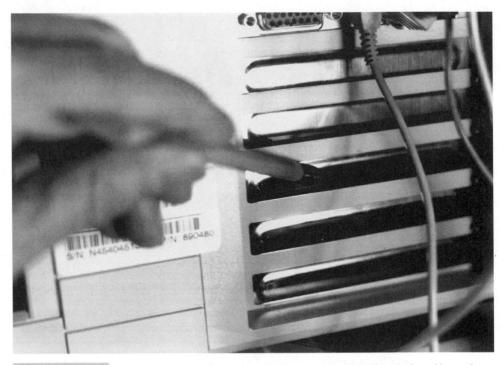

**FIGURE 9-21**    **These opening slots are used for air-cooling circulation. Keep them clean and do not cover them.**

**Booting from the disk drive**  Your PC cannot operate unless it has an operating system. All modern PCs obtain this program from the disk drive, and then loads it into the RAM. Thus, when you turn on the PC, it can automatically load the operating system program very simply. It then runs the computer self-test and looks at all of the drives for an operating system or programs.

The reason that the hard-drive operating program is used and not permanently held inside the PC on the RAM is because the program can be easily upgraded. Also, if the system program has some errors or glitches, and they usually do, this can easily be corrected. So, it is much simpler and cost-effective (cheaper) to load in a new program from a CD-ROM or floppy disk than to replace the RAM chips. In more sophisticated PCs, you can even choose one of several operating systems each time you turn on the computer.

**PC operating system**  If your PC did not have an operating system, it would be much more troublesome and time-consuming to operate the various program you want to run. If your PC did not have this common platform, it would be next to impossible to use any of your software programs.

Now that your PC has successfully loaded the operating program and all systems are go, you can now use the mouse to point to the Start button and click on the program you want to use. The screen will look like Fig. 9-22.

**The random-access memory (RAM)**   Your computer must have RAM IC chips before it can operate and run a program or do useful work. When the PC is first turned on, the RAM contains no information (it is erased when the computer is shut down).

> Some newer and expensive PCs have advanced RAMs that hold the memory after PC is turned off. This is the reason the PC has to take a little time to load a program from the hard drive into the RAM for the PC to operate. The various documents and databases are then loaded into the RAM, usually for a short period, before your software can be manipulated by the CPU.

All computers, whether laptops, PCs, or huge super models, use the same digital type of information, commonly called *machine language*. This language is in the form of Binary numbers. This data is all in 0s and 1s, highs and lows, or offs and ons. This simple 1 and 0 scheme can represent millions of words and numbers. Some of the first crude computers used simple off and on switches to install the program. It was a very slow way to go for the computer programmer. These same 1s and 0s are written to and read off the disk drives and memory chips in your PC.

For more information on basic electronics and how transistors and ICs work, refer to Chapter 1. Chapter 1 also includes more details on binary.

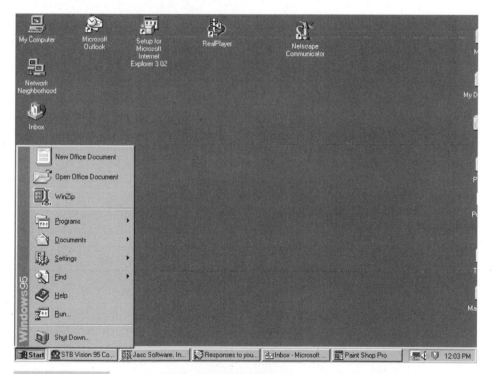

**FIGURE 9-22**     The screen indicates that the PC program is now loaded OK and you can click on the start button at lower left of screen.

**Why a RAM cache is needed**   All types of computers have RAM, which vary in speed and cost. The RAM cache speed determines how fast the data can move from the RAM cache to the CPU. The RAM cache has at least 64KB (kilobytes of memory).

With no RAM cache, the CPU might not have any work to process during many clock cycles while waiting for more data. With the cache, the data will be quickly available whenever the CPU is ready to process it.

As noted earlier in this chapter, the CPU is the heart of the system. The speed that the CPU clock oscillator can run determines how fast the programs can be processed by your PC. As an example, an Intel Pentium microprocessor might have a clock speed of 100 MHz to 400 MHz. One MHz is equal to one million cycles or bytes processed every second. Of course, you also need lots of fast RAM chips and a hard drive with lots of memory space to process lots of data at a rapid clip.

**Disk storage operation**   The computer disks, usually referred to as 3½-inch floppies, are a very economical form of permanent data storage. On these disks, you can store hundreds of kilobytes of data. All of these disk drives put the 1s and 0s data binary language on their surface in various schemes. This data information is written to and read from the disks in magnetic form.

There are various ways in which the disk can be organized with this data or formatted. One common method is Microsoft's MS-DOS disk system.

When you install a new floppy disk, you will have to format it. Formatting is a technique to "tell" the data how it should be put down on the disk. Actually, magnetic codes are inlayed within the surface of the disk. As shown in Fig. 9-23, the disk is divided into sections or circular tracks. This is done so that the read and write heads can quickly move all over the disk surface and find or access the requested information. If you no longer need the information on the disk, it can be erased and used over again many times. To do this, merely reformat the disk again.

**Floppy disk-drive operation**   The floppy disk-drive system is very cost effective and efficient computer storage device. Of course, it is slow, compared to other type drives, but it is a most useful item.

Figure 9-24 shows the location of the 3½-inch floppy drive inside the case. These floppy disks can be taken from one PC and read in another. The 3½-inch floppy is very compact,

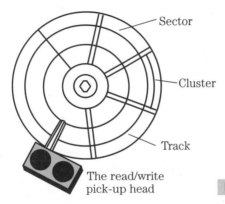

Sector

Cluster

Track

The read/write
pick-up head

**FIGURE 9-23**   **How the PC disk is formatted.**

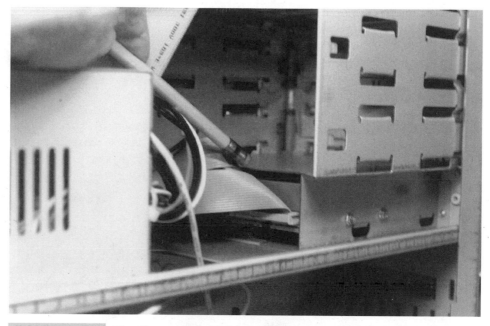

**FIGURE 9-24**    **The floppy-disk drive inside the PC.**

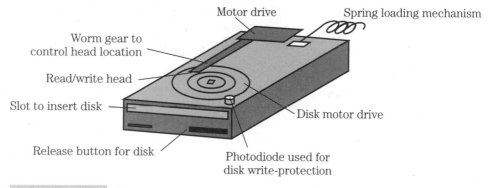

**FIGURE 9-25**    **The 3½" floppy drive.**

inexpensive, and holds a lot of information. A mechanical layout drawing of the 3½-inch floppy drive is shown in Fig. 9-25.

A floppy disk is shown being properly installed in a PC in Fig. 9-26. The floppy disk write-protection tab is shown in Fig. 9-27. Do not let dust and dirt get into the slot where the floppy disk is installed. Keep the floppy disks clean and at room temperature. Do not place the floppy disks close to a magnetic field, such as an audio speaker or a magnet. Also, keep the disks away from any electronic devices that can emit strong electrical fields, which can erase the information data on the disks. You might want to keep these floppies in a metal box or in a steel file cabinet.

FIGURE 9-26    A floppy disk being installed in a loading slot.

FIGURE 9-27    The floppy slide switch is used to save data that you do not want to be erased. You cannot write to the disk with the switch in this position.

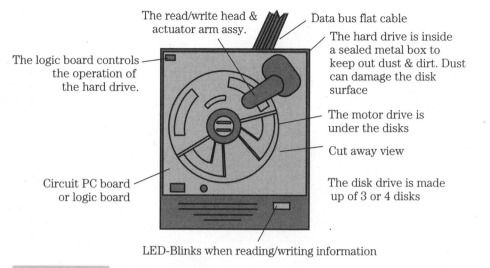

The read/write head &
actuator arm assy.

Data bus flat cable

The hard drive is inside
a sealed metal box to
keep out dust & dirt. Dust
can damage the disk
surface

The logic board controls
the operation of
the hard drive.

The motor drive is
under the disks

Cut away view

Circuit PC board
or logic board

The disk drive is made
up of 3 or 4 disks

LED-Blinks when reading/writing information

**FIGURE 9-28    The hard drive.**

**Hard disk-drive operation**  The computer hard drive is a hard-working component. Millions and billions of bytes can be stored and retrieved from these disks that turn at very fast revolutions. When you turn on your PC, the various noises you hear are the hard drive getting information, loading the programs, and checking various operating procedures. The hard-drive read/write heads must do their work with precision because the disk spins very high revolutions and pick up very small information bits on its surface. The hard drive is an electrical and mechanical device and is always spinning when the computer is turned on. For as much as it has to do, it does not fail often. Also, in the past year or so the price has gone down and the speed and data capacity has shot way up. Most PCs now have hard drives more than 2-Gb capacity. The physical size has also gotten smaller, too. Many hard-drive units will fit into the space of a 3½-inch floppy drive. Figure 9-28 shows a hard drive.

With Windows 95 program installed in your PC, when you delete a file it goes to a holding place on the hard drive, called the *recycle bin*. So, if you delete some program or work by mistake, you can go back to the recycle bin and retrieve it. This can be a lifesaver as I have discovered a few times. However, the hard drive can become full, and you need to empty the bin from time to time to make more space available.

**The operations of a CD-ROM drive**  The CD-ROM in your computer is very much the same as an audio CD used in stereo systems. In fact, if you have an audio board and amplifier in your PC, you can listen to CD music as you work at your PC. As with a music CD, the CD-ROM can store a lot of data in a very small space. The CD system retrieves the recorded data with a beam of laser light that's very finely focused. Figure 9-29 shows an inside view of the various components that make the CD-ROM operational. Figure 9-30 shows the location of the CD-ROM drive unit mounted inside the PC computer case.

Figure 9-31 shows the proper placement of the CD-ROM into the slide drawer. Always install the disk with the label side up. And always keep the slide drawer clean of dirt and dust. If the ROM program will not load or play the program properly, be sure that the disk is clean. Be sure any fingerprints are cleaned off the CD surface. If the mirrored surface is

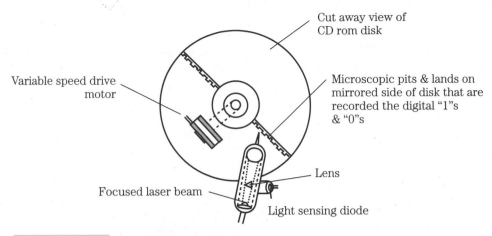

Cut away view of
CD rom disk

Variable speed drive
motor

Microscopic pits & lands on
mirrored side of disk that are
recorded the digital "1"s
& "0"s

Focused laser beam

Lens

Light sensing diode

**FIGURE 9-29    The CD-ROM drive.**

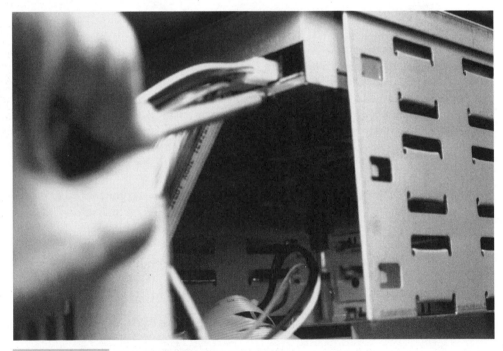

**FIGURE 9-30    The CD-ROM in the PC.**

dirty, the laser beam cannot properly read the digital information contained in the pits and lands of the CD's surface.

**The inside workings of the mouse**    The mouse (Fig. 9-32) is used on your PC as a pointing device. With the mouse, you can manipulate programs quickly without having to type in instructions on the keyboard. Other screen-control devices are the joystick, light pen, or a special touch screen. Now see how moving the mouse around can cause PC screen pointer action.

Figure 9-33 shows the bottom view of a typical mechanical mouse. When you push the mouse around on a flat surface pad, the rubber ball on the bottom controls two roller

**FIGURE 9-31**   Proper installation of the CD-ROM.

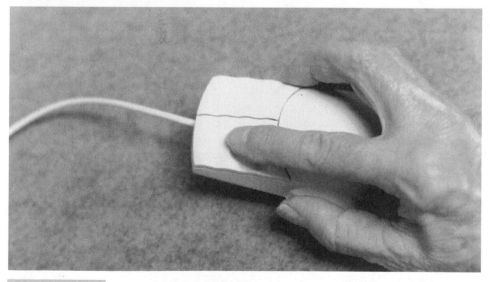

**FIGURE 9-32**   A top view of the mouse.

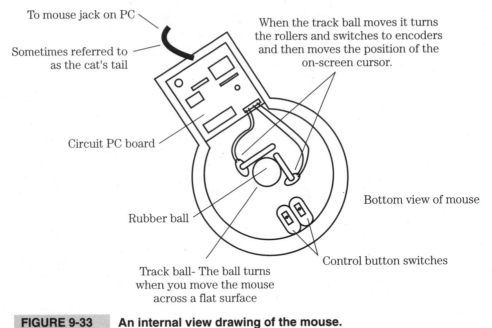

To mouse jack on PC

Sometimes referred to
as the cat's tail

When the track ball moves it turns
the rollers and switches to encoders
and then moves the position of the
on-screen cursor.

Circuit PC board

Bottom view of mouse

Rubber ball

Control button switches

Track ball- The ball turns
when you move the mouse
across a flat surface

**FIGURE 9-33**    **An internal view drawing of the mouse.**

shafts that have rotary switch contacts on one end. The switch contacts for the vertical
and horizontal shaft movements give the mouse's direction via electronic signals. These
switch encoded signals are then sent to the PC circuit board, where they are sent by the
cable wires to your PC's software, which controls screen pointing.

The two control button switches are the ones you click to perform various software program operation.

**Some probable mouse troubles and solutions**    If the mouse is not moving the screen
pointer (or is moving it erratically), take the mouse apart and check it. A bottom view of
the mouse is shown in Fig. 9-34. The mouse might have a screw in the bottom to remove
or clip tabs to unsnap the bottom cover. Some are sealed and cannot be taken apart.

Check the rubber ball and be sure it is clean and is able to roll around freely. The ball
should have good contact with the roller shafts and be able to rotate the switches. Clean all
of the switch contacts and the ball with a good electronic contact cleaner. Also check for
loose or broken cable leads and poor connections. Clean the mouse port plug-in connection to the PC. It is always best to use a special mouse pad to grip the ball properly.

On some PCs, when you replace the mouse with a new one, you might have to repro-
gram or reboot the computer before the mouse will operate. This information should
come packed with the replacement mouse.

**PC keyboard operation**    Some of the new PC keyboards today have a built-in trackball
and even a scanner to load printed document pages directly into your PC. When buying a

new PC, always try out the keyboard and look for one that is easy and comfortable to type on. You might want to consider some new designs of keyboards. Now see how the keyboard works and how information from the keys is delivered to your computer.

Figure 9-35 shows a side view of one contact key that is commonly in use today. The key has a soft center material that is squeezed down by your finger as you touch the key cap to type. This material then pushes down a plastic membrane sheet that has a set of flat metal

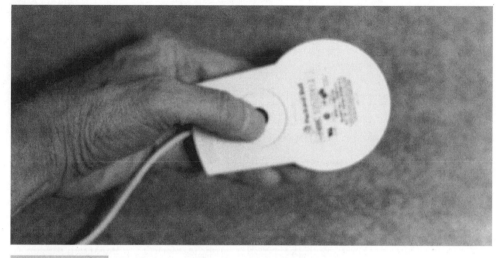

**FIGURE 9-34**    The bottom of the mouse and track ball.

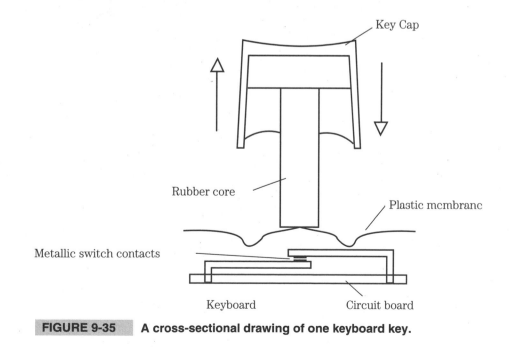

**FIGURE 9-35**    A cross-sectional drawing of one keyboard key.

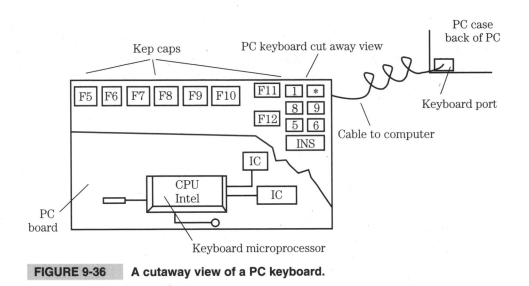

**FIGURE 9-36    A cutaway view of a PC keyboard.**

contacts that are switched on momentarily. These are connected to a PC board that is located on the bottom of the keyboard. The current from these contacts is routed to a micro-computer chip located on the PC board. This microprocessor generates a scan code, according to which key is being pressed down. Figure 9-36 shows a partial keyboard.

The scan code from the keyboard microprocessor is then translated into an ASCII code via a cable to the keyboard port on your PC. This code is then used by the PC to produce the various characters on your monitor screen.

**Modem operation for your PC**   For you to use the Internet, your computer must have a modem (*mo*dulation and *dem*odulation). You can think of the modem as a "go-between" of digital and analog signals. The unit changes digital data to a varying electronic frequency waveform. When the modem is receiving a signal from the phone line, it demodulates the analog signal and changes it to a digital form. In essence, the modem makes it possible for two or more computers to "talk" or exchange data between them, as requested.

The modem is needed to send information back and forth over a phone line and can be an external or internal device. If you do not have a modem, you can add an external one to your PC so that you can surf the Internet. A drawing of an external modem is shown in Fig. 9-37. The call-outs of the LEDs (light-emitting-diodes) located on front of the modem box give you an indication of how your modem is operating.

As covered previously, your PC is a digital operating system that manipulates data by electronic off/on switches. This is called *binary data operation*.

The telephone uses analog transmission design to convey voice, tones, and other audio sounds. It was many years after the fast telephone system was in place that digital electronic devices were created. Also, digital switching does not work well over copper phone lines. The telephone sounds are transmitted as a continuous smooth current that always varies in frequency and amplitude or strength. The analog or audio signal is shown in Fig. 9-37, as it would appear on an oscilloscope.

The modem speed (how fast data can be transmitted back and forth) is measured in baud (frequency changes that happen every second) or bits per second, which most modems are rated at today. A common bit rate is 33,600 bits per second and for faster data transfer, 56,700 bits per second are used. The faster, the quicker you can acquire or upload information.

If you have an intermittent operating modem that is internal in your PC, you can turn the computer off or re-boot it to start operation again. At times, because of various power-line glitches, etc., your modem might lock up and not transmit or receive data. If you unplug the power from it to turn it off for a minute or two, then power it back up, it should start operating again.

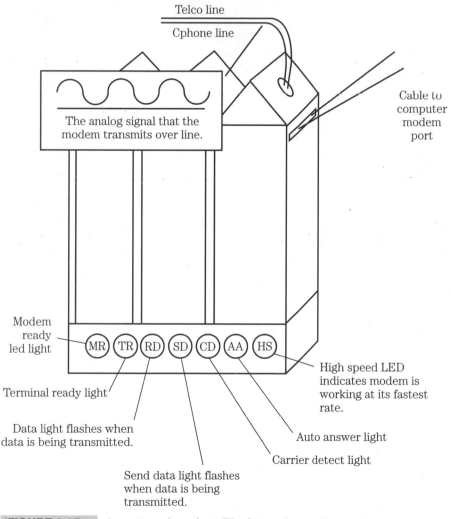

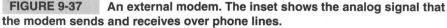

**FIGURE 9-37** **An external modem. The inset shows the analog signal that the modem sends and receives over phone lines.**

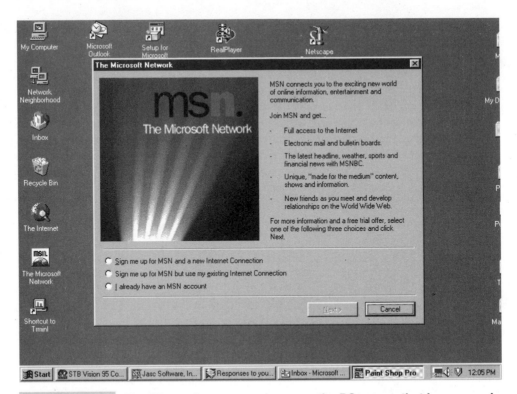

**FIGURE 9-38** The Microsoft program shown on the PC screen that is now ready for using the Microsoft network.

**Internet access notes** If you want to cruise the Internet, you have to obtain access and an address. It is best to obtain local phone-line access to avoid long-distance charges. Also, DirecTV will be featuring Internet downloading from their dish sometime in late 1998. For more access information, call the following phone numbers:

- American Online (800-827-6364)
- CompuServe (800-848-8199)
- Delphi (800-695-4005)
- GEnie (800-638-9636)
- Prodigy (800-776-3449)
- MSN MicroSoft Network is shown in Fig. 9-38 on the PC monitor.

**The computer monitor operation** All PCs have a color monitor to properly display all of the many varied graphic programs, CD-ROMs, and Internet services. The color monitor needs high resolution (the higher the pixel count the better) to obtain the best video pictures. The most common picture tube uses the dot-matrix shadow mask system and also the very high-quality Sony Trinitron picture-tube system.

A cut-away view of a typical PC monitor is shown in Fig. 9-39. The best monitors use the Super Variable Graphics Array (SVGA) for the video display. The SVGA technique uses an analog signal, which is then converted into digital information with different voltage levels. Then, the dot or pixel brightness level is varied on the monitor screen.

For more information on how a monitor operates, refer back to Chapter 4, which explains color TV operation and the color picture tube design.

**Cleaning the monitor**   Because of the high voltage used to operate the picture tube, dirt and dust tend to be drawn to and stick onto the CRT screen faceplate and into the case. Use a little Windex on a cloth to wipe the screen and the monitor case. Be sure that all of the air vents on the back and the bottom of the monitor case are clean and free of any obstructions. You can use a vacuum cleaner to pull out the dust and dirt.

⚠️ Do not spray any cleaner or water into the vents of the monitor or the PC housing vents.

# Basic Microprocessor System Overview

Think of a system with a keyboard and a display as somewhat like a pocket calculator. When a key is pressed, the corresponding number or letter will appear on the display. This system is a natural application for a microprocessor; in many ways, it is similar to a mini-computer.

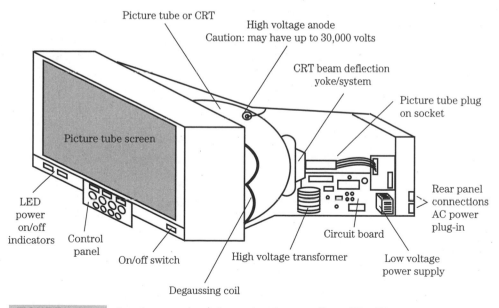

**FIGURE 9-39    A cut-away view of a computer monitor with all key component location call-outs.**

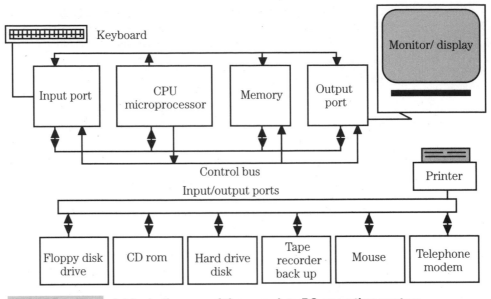

**FIGURE 9-40**    A block diagram of the complete PC operating system.

Figure 9-40 brings all of the PC functions together for an operations system overview. The microprocessor (CPU) is the "brains" of the system. It contains all of the logic to recognize and execute the list of instructions (programs). The memory stores the program and can also store data. Other program and storage devices are the floppy drive, CD-ROM drive, and hard drive. The printer and modem are extra devices.

The microprocessor needs to exchange information with the keyboard, monitor display, printer, and various other disk drives. The input port, from which the CPU can read data, connects the CPU to the keyboard. The output ports, to which the microprocessor can send data, connects the CPU to the monitor display, printer, modem and etc.

The blocks within the PC computer are interconnected by several buses. A bus is a group of wires that connects the devices in the system in parallel. The microprocessor uses the address bus to select memory locations or input and output ports. You can think of the addresses as your home mail box because they identify which locations to put information into or take information from.

Once the microprocessor selects a particular location via the address bus, it transfers the data via the data bus. Information can travel from the microprocessor to the memory or an output port, or from an input port or memory to the microprocessor. Notice that the microprocessor is involved in all data transfers. Data usually does not go directly from one port to another, or from the memory to a port.

Another bus is called the *control bus*. It is a group of data signals that are used by the microprocessor to inform memory and input/output (I/O) devices that it is ready to perform a data transfer. Some signals in the control bus allow I/O or memory devices to make special requests from the microprocessor.

A single digit of binary information (1 or a 0) is called a *bit* (a contraction of *bi*nary digi*t*). One digital signal (high or low) carries one bit of information. Microprocessors handle data, not as individual bits, but as groups of bits, called *words*. The most common modern microprocessors use 16- and 32-bit words, which are called *bytes*. These microprocessors are then called 16- and 32-bit processors. For a 16-bit processor, *byte* and *word* are often used interchangeably. Keep in mind that *word* can be used for any group of 8, 16, or 32 bits.

## PROGRAMS

A program is required by the system to perform the desired task. In the following examples, some of the instructions are required:

- Read data from the keyboard.
- Write data to the monitor display.
- Read and write data from the hard drive.
- Read and write data from the floppy disk.
- Read program from a CD-ROM.
- Read and write data from internal memory chips.
- Write data to the printer.
- Request data from the Internet and download the data to the PC memory for future use.
- Perform computer testing routines.
- Perform data backup to tape.
- Repeat requested programs.

For the microprocessor to perform a task from a list of instructions (program), these instructions must be translated into a code that the microprocessor can understand. These codes are then stored in the computer system's memory. The microprocessor begins by reading the first coded instructions from its memory bank. The microprocessor decodes the meaning of the instructions and performs the indicated operation. The microprocessor then reads the instructions from the next location in memory and performs the corresponding operation. This process is repeated, one memory location after another.

Certain instructions cause the microprocessor to jump out of sequence to another memory location for the next instruction. The program can, therefore, direct the microprocessor to return to a previous instruction in the program, creating a loop that is repeatedly executed. This enables system operations, which must be repeated many times to be performed by a relatively short program.

## COMPUTER PERIPHERALS

A complete microprocessor (CPU) system, including the CPU, memory, hard drive, CD-ROM, floppy disk drive, and input/output ports is called a *computer*. The devices connected to the input/output ports (the monitor display, keyboard, mouse, and modem) are called *peripherals* or *input/output (I/O) devices*. The peripherals are the system's interface with the user. They can also connect the microcomputer to other equipment. Storage devices, such as tape drives, fax boards, printers, and scanners are also referred to as *peripherals*.

This should now give you a good overview of how everything in your PC operates and is all tied together. Now look at some PC problems you might encounter and their solutions, plus some preventive maintenance procedures that you might wish to do.

# Tips to Keep Your PC in Top-Notch Condition

It is always worth your time and effort to keep any electronic device clean of dirt and dust.

## SLIDING CD DRAWER

Keep the CD drawer clear of dust and also wipe the CD disc clean often. Be sure that the CD drawer slides in and out easily. Push the Eject button and note how it slides in and out. If you observe any erratic motion, the slide runners might need cleaning.

## FLOPPY DRIVES

The floppy-disk drives should be cleaned often. These drives will accumulate dust and magnetic oxides that are deposited after a long time of disk usage. This could result in random disk read/write errors that might appear as "Failure to Read Drive A," etc.

If you run your floppy drive often, you should invest in a cleaning drive kit. These cleaning kits can be used quite a few cleaning cycles. One type of kit is dry and uses no fluid, but is more of an abrasive cleaning disk. The other type is a wet-variety kit; you apply a cleaning fluid to the fabric disk. I would not advise running either of these cleaning disks more than 20 or 30 seconds for each cleaning operation.

*Problem*   There is no action on the PC at all. LED lights do not come on, drives do not work and no noise from hard drive or blower fan.
*Probable causes*

1 Be sure that the ac outlet where your PC is plugged has power. Check it with a lamp or voltmeter.
2 Check for loose connection between the computer and power cord plug-on socket.
3 If your PC is plugged into a surge suppressor, the suppressor might be bad. Check by plugging your PC directly into the wall socket.
4 Check other equipment, such as the printer and monitor, for proper power. If they are OK, switch the power cords between them and the computer.
5 Check the power on/off switch for proper action.

*Problem*   The computer produces error messages when booting up, starting Windows 95, or trying to start other applications.
*Probable causes*   For a bad command or file name, check the following:

1 Some PCs include a master CD-ROM or you can order one. Use the master CD-ROM to restore the hard-disk drive to its factory configuration. To do this, just insert the master CD into the CD-ROM drive drawer and follow the installation directions.
2 You might have a menu come up with a Restore/Recovery option. Click on it with the mouse. Under the PBFIX option, you can then restore Navigator or Windows 95, with the option you want.

*Problem*   No picture is on the monitor screen. The computer appears to operate. The lights come on, the fan hums, the computer accesses the disk drives, but you do not see a picture on the screen. A power-on light might be on the monitor.

*Probable causes*   Be sure that the monitor is getting ac power. Use the same checks as you made for verifying power to the computer.

If the monitor is receiving power, check the cable connections between the monitor and the computer. With both PC and monitor power turned off, disconnect the monitor cable from the computer. Be sure that no pins are bent or damaged. Usually, these plugs have pins missing, which is normal. If the monitor cable can be unplugged at the monitor, check the connections on this plug, also. Clean the pins in the plug and socket connectors.

If you still do not see a picture, turn the Contrast and Brightness controls clockwise to their maximum range. The screen should now turn light gray and you will see a raster. The raster is the many lines that produces an image on your monitor. If you can see the raster, it usually indicates that the picture tube is good, and the problem will be in the video circuits inside the monitor. You can usually have the monitor repaired at a video electronics repair center.

*Problem*   Computer halts during start up. When the computer stops, it displays the message "Non-System Disk or Disk Error."

*Probable cause*

1  Check the floppy drive and be sure that no disk has been inserted. If a disk is present, remove it, and press Enter to start the computer. The message "Starting Windows 95" should now be displayed.
2  Insert the multimedia master CD (on multimedia systems) or the recovery disk in the appropriate drive. Restart the computer using the CTRL-ALT-DEL key combination. If you are using the master CD, choose the menu option to go to a DOS prompt.

Type the command:

```
SYS C: [Enter]
```

After a few seconds, this message should appear on the PC monitor screen:

```
SYSTEM TRANSFERRED
```

Remove the CD and restart the computer using the CTRL-ALT-DEL keys. This should now clear your disk error.

## DIAGNOSTIC SOFTWARE DISKS

Many of your PC problems will not be caused by electronic or mechanical failures, but problems because of defective software programs and various glitches and other virus contaminations. Software disks are available to clear up these problems. This software is used to determine and correct problems in a system that can be rebooted. Some of these disks can tell what components are in the PC, their status, and how they are configured. If the CPU, hard drive, and some of the RAM chips are OK, then these software disks can diagnose other problems in your PC.

## DISK UTILITIES FOR HARD DRIVE TESTING

You can use a test utility disk when the PC hard drive cannot be accessed, but there cannot be any PC failure or when the hard drive will not boot up, but can be booted from the floppy drive. These fixed disk drive utilities are used to test, repair, and recover the data from your PC hard drive.

## UTILITY FOR THE FLOPPY DRIVE

These utilities can be used when your floppy drive gives an error, but is not the floppy's fault. These utility floppy disks will generally clean, test, and realign the drives. The utility floppy will also perform a head-cleaning routine and will move the floppy heads across the complete surface of the cleaning disk. And an added feature if there is still a problem after cleaning, the utility disk can run a test and probably find the floppy fault.

## PRESCRIPTION FOR CURING COMPUTER VIRUS

Your computer can become contaminated with a virus in many ways. It can be introduced from a floppy-disk software that was used in another PC that you borrowed, software that you purchased or data that was downloaded from an on-line source. So, to be on the safe side, you might want to load in a virus-catching software program.

As an example, with Windows 95, you might start having "odd-ball" messages show up on your monitor such as abort, retry, ignore, or master boot record has been modified. To find and correct these virus problems, you want to invest in a Dr. Soloman's Anti-Virus Toolkit. Call (617) 273-7400 for more information or to order. The best Anti-Virus programs are loaded into your new PC before it's used. What the program does is just keep checking each disk, you use or data downloaded and it catches the virus before it can damage your whole PC system.

## ANOTHER FIX-IT DISK

Besides the anti-virus programs, many other software programs locate and repair computer problems. In fact, they are almost a must-have if you are running the Windows 95 system. This system is very complicated and you just cannot remember all of the unusual messages that jump out at you at various times.

One good repair software kit is the "PC Handyman" that is available from Symantec. You can order or find out more about this kit by calling 800-441-7234. This program runs in the background of other programs and when "odd-ball" problems occur, the PC Handyman should recognize the fault and fix it.

# In Summation

Always be cautious when working on your PC and monitor because you can get shocked. Unplug the power from the ac wall socket for the sake of safety. Keep the keyboard, case, and monitor clean. Keep the floppy disks clean of dust and dirt and wipe away smudges from the CD-ROMs. And plug your PC into a UPS.

# 10

# PRINTERS, COPIERS, AND
# FAX MACHINE OPERATIONS

# Daisywheel Printer Operation

The daisywheel printer system is used in *personal word processors (PWPs)* and electric typewriters. A drawing of the daisywheel layout is shown in Fig. 10-1. A photo with two daisywheels is shown in Fig. 10-2. Figure 10-3 shows the actual placement of the daisywheel in a PWP printer. A typical PWP is illustrated in Fig. 10-4.

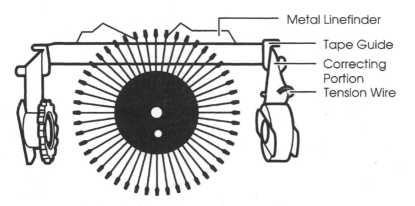

Metal Linefinder
Tape Guide
Correcting Portion
Tension Wire

**FIGURE 10-1**    The "daisy" wheel found on some printers and typewriters.

**FIGURE 10-2**    A close-up view of two "daisy" wheels.

**FIGURE 10-3**     A "daisy" wheel being installed on an Olivette.

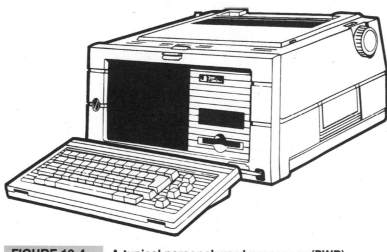

**FIGURE 10-4**     A typical personal word processor (PWP).

The daisywheel printer is an all-electronic machine that prints faster and is quieter than a typewriter. These printers are faster because they print, bi-directionally, to both left and right on alternating lines at 20 characters per second. However, these daisywheel printers work at a much slower speed than the other printers covered later in this chapter.

The heavy-duty printwheel (Fig. 10-1) consists of spokes or petals. Each spoke contains a unique character. The printwheel will actually rotate very fast to the left or right, depending

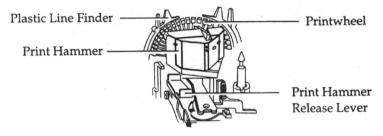

Plastic Line Finder — Printwheel

Print Hammer — Print Hammer Release Lever

**FIGURE 10-5** **The print wheel and print hammer that transfers images on the daisy wheel to the paper by-way of a ribbon.**

upon the character that has been typed from the keyboard, or with a printer, the wheel is controlled by the PC. The printwheel and print hammer is shown in Fig. 10-5. An electronic digital IC interface that is controlled by the keyboard tells the printwheel when and where to spin and then activates the hammer solenoid to strike the printwheel character and make its mark. These wheels come in various styles and sizes that you can interchange to produce different looking documents. These wheels come in "pitches" of 10, 12, or 15 characters per inch.

These printwheel printers use ribbons and can be made of fabric, carbon, or a correctable ribbon. Some machines even have a correcting tape when they are used as a typewriter. Most of these machines will print documents with either a ragged right margin (uneven) or a justified margin (perfect straight right margin).

# Daisywheel Printer Tips

## DATADISKS

Although the DataDisks are not fragile, certain precautions should be followed.

- Do not place the DataDisk near any magnetic object.
- Do not expose the DataDisk to temperature extremes.
- Do not bend the DataDisk.
- Do not store in any power cord storage compartment because they are usually close to an electromagnetic device, such as a power transformer.

## KEYBOARDS

To clean covers or keyboards, sponge it off with a mild ammonia or soap solution. Do not use household cleaners containing chlorinated components.

## PRINTWHEEL

To remove any residue from the printwheel, dip the characters wheel edge into a small container of ethyl or isopropyl alcohol (rubbing alcohol) and wipe with a clean dry cloth. Do not soak the printwheel.

## PLATEN CLEANING

Wipe the platen surface off with a mild soapy solution. Do not use household cleaners containing chlorinated compounds.

## MONITOR SCREEN

The monitor screen should be cleaned with the power turned off. Dust with a dry, soft cloth or use a good-quality CRT screencleaning kit that will neutralize static and will not streak or scratch the monitor screen.

# Checks for PWP Machines

If your PWP does not function properly, then perform the following checks:

- Check for proper position of the correction tape spool.
- Does the ribbon cassette cartridge need to be replaced?
- Is the top lid closed tightly?
- Does the correcting tape need to be replaced?
- Has the print carrier been released?
- Has the printwheel been installed correctly?
- Has the printwheel been installed?
- Has an object fallen into the carriage and jammed its operation?
- Has the print hammer device been positioned correctly?
- Is the monitor screen dim or blank? Try adjusting the contrast and brightness controls.

# How the Ink-Jet (Bubble) Printer Works

Now, find out how the ink-jet printer works. The ink-jet printer is also referred to as a *bubble-jet printer*, which will become obvious as the system is explained. The Cannon Model 1000 is used for the ink-jet system operation (Fig. 10-6). When connected to your PC, the Model 1000 does not only make print documents, but can make copies, faxes, and scans. The ink-jet printers are very compact, lightweight, and have about the same print resolution as a laser printer. Compared to the standard typewriter and daisy-wheel machines, the ink-jet is very quiet during operation. The ink-jet machine prints the paper by squirting small droplets of ink out of the print head nozzles. Figure 10-7 shows the print head and ink cartridge being replaced on the Cannon Model 1000 multi-pass machine. Figure 10-8 gives you details for replacing the ink-cartridge/print-head assembly.

An ink-jet printer contains four major blocks (Fig. 10-9). These blocks include the print head, paper handler, carriage transport, and electronics logic control board. A motor starts the print head moving along a track and IC printer circuits send a voltage pulse to each head nozzle, which then leaves the proper mark on the paper.

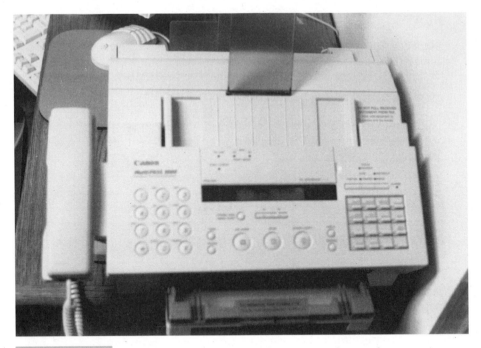

**FIGURE 10-6**    The Canon model 1000 printer, copy, fax, and scan multimachine.

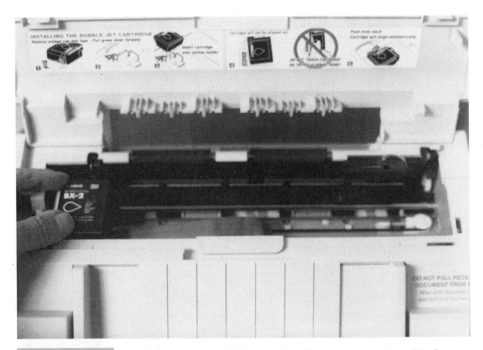

**FIGURE 10-7**    The print head and ink cartridge being replaced on the Canon model 1000 machine.

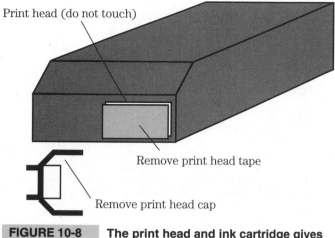

Print head (do not touch)

Remove print head tape

Remove print head cap

**FIGURE 10-8** **The print head and ink cartridge gives you the ink cartridge replacement details.**

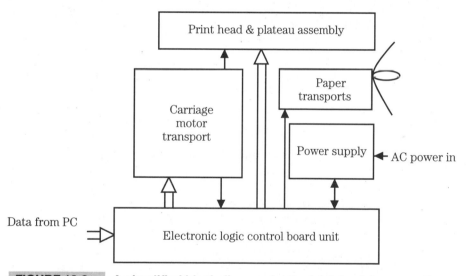

Print head & plateau assembly

Carriage motor transport

Paper transports

Power supply ← AC power in

Data from PC ⇒

Electronic logic control board unit

**FIGURE 10-9** A simplified block diagram for the ink-jet printer operations.

## PRINT CARTRIDGE AND NOZZLES OPERATION

The printer cartridge contains many ink-filled chambers that feed into each ink-jet nozzle. Figure 10-10 shows the print head, which contains many fine nozzles, whose diameter is smaller than a human hair. Each ink chamber and hole has a thin resistor (or heating element) that is fed a controlled electrical pulse from the logic control board at the proper time to heat the ink to more than 900 degrees for a few millionths of a second.

Thus, the ink is heated to form a bubble vapor. When the heated bubble expands, it is pushed through the hole. The pressure of the vapor bubble shoots the ink droplet onto the paper. The print character is then formed by an array of these droplets pushed through the

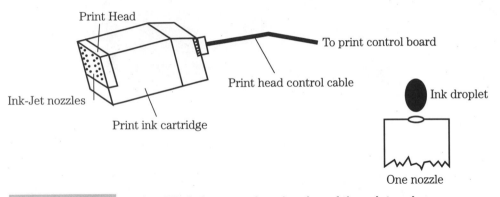

**A simplified close-up view drawing of the print and ink-jet nozzles.**

micro holes. The more and finer nozzle holes in the print head, the better the printer resolution or sharpness will be.

## INK-JET HEAD PROBLEMS

These ink-jet printers will print on many different types of paper surfaces. They have a higher printer speed than daisy-wheel printers and are very quiet. If your machine is not printing properly, be sure that you are using quality paper for ink-jet printers. Porous paper will absorb the ink and make your printed documents look faded and dull.

## INK-JET PRINTER PROBLEMS

Now take a look at a few common printer problems. Many print problems are caused by a defective print head or it has run out of ink. The ink-jet machines use a print head/cartridge combination module that is very easy to snap in and out. You might see some of the print characters partially missing and this is probably caused by some of the nozzles being plugged up in the print head. In this case, you must replace the head cartridge. Other printing problems can be caused by a defective IC in the logic control unit or head-driver electronics. Try cleaning all cable plug pins and push-on connectors. If the machine will not print at all, check the status lights and see if it is on line. Also check the interface cables and connector plugs from your computer. If the print head moves back-and-forth, but does not print, suspect a defective print head, out-of-ink cartridge, defective ribbon cable, or dirty or broken connectors to the print head. Then recheck and be sure that the print head is installed correctly.

## PAPER-HANDLING PROBLEMS AND CHECKS

Printers use two different types of paper-feed systems. These are friction feed, like a typewriter uses, and the tractor feed, which requires special paper with notches on each edge of the paper (sometimes referred to as *computer paper* or a *continuous-feed paper*). Paper-feed problems can be caused by paper not installed properly, wrong type of paper, or mechanical problems. Be sure that the correct paper is being used and that it is installed properly. Then check and clean the platen and pressure rollers. Check and clean any drive gears and chain or belt drives. Be sure that the gears mesh and move freely. The drive feed motor might be defective or a fault might have developed in the logic IC control board.

## PRINT-HEAD CARRIAGE ASSEMBLY PROBLEMS

The ink-jet printers move the print-head from left to right and back again on a rail to print across the page surface. Figure 10-11 shows the BX-2 print head cartridge on the left side; as it prints, it moves to the right side on a rail guide and is pulled by a belt or chain drive powered by the carriage motor.

If the print head does not move across the carriage rod or moves in jerks and does not position itself at the left side when the printer is turned on, it probably has a mechanical problem. This problem could be a loose or broken carriage belt, chain, drive gear, or pulley, and possibly a faulty carriage motor. If the belt is loose or broken, you will need to replace it. If your machine uses a carriage chain or gears, they might only need to be adjusted or cleaned. Also, check, clean, and tighten all cable plug-in pin connections to the motor and control PC boards.

**FIGURE 10-11**    The BX-2 print head is shown on the left side and it moves to the right side on a carriage rod to print lines across the paper.

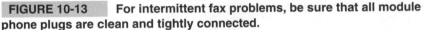

**FIGURE 10-12**    A surge-protection device for your printer that plugs into an ac outlet. It also has connections to plug in phone and fax equipment for phone-line protection.

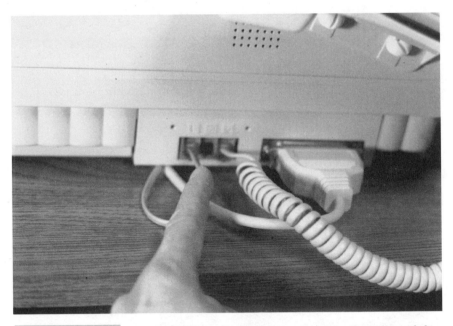

**FIGURE 10-13**    For intermittent fax problems, be sure that all module phone plugs are clean and tightly connected.

If the belt, cable connections, and motor checks out OK, then the problem will be in the electronics logic control PC board's drive circuits, the optical encoder, or a power supply problem. Always check the power supply for correct dc voltages when you have a print head, carriage transport, control or drive circuit, and paper-feed problems. To prevent ac line surge damage, plug your printer into a protection device (Fig. 10-12). On fax machines, always check and clean the phone module plugs that are shown in Fig. 10-13.

## SOME MULTI-PASS TROUBLESHOOTING TIPS

This section for the Cannon MultiPass machine looks at software problems.

### Printout is wrong

- Check and see that the cable connections are clean and tightly secure.
- If you are printing in DOS, check that the printer control mode matches the driver that you selected.

**Print job vanishes**   If you have a print job that vanishes or are printing garbage, the problem might be because another Windows application on your PC is trying to communicate with the printer port of the MultiPass Server it is using. This conflict will generally result in strange printed garbage.

> The Preference dialog box includes an option to use the Desktop Manager Spooling. You do not need to change the Print Manager or Print Spooler settings in Windows and the print settings for other Windows applications or other printer drivers that are not affected.

## BUBBLE-JET PRINT JOBS DISAPPEAR UNDER WINDOWS

The MultiPass software includes a driver named MPCOMM DLL. This driver adds to the standard Windows communications driver named COMM.DRV. Under Windows 3.1, MPCOMM.DLL serves to improve bubble-jet printing performance and ensures that the MultiPass software does not interfere with any serial communications that is set properly.

## CHARACTERS ON SCREEN DO NOT MATCH PRINTED CHARACTERS

Many graphics characters and special symbols are produced by different ASCII codes on each make of computer and printer. You will need to reset the correct character table and printer control codes if you encounter this problem.

## PRINTOUT DOES NOT MATCH PAPER SIZE

- Be sure that the paper size you selected in the software matches the paper loaded into the unit.
- Be sure that the width of the paper on which you are printing matches the width indicated by your software so there is always paper between the print head and the platen. The print head might be damaged if it prints directly on the platen. If the print head prints on the platen, feed a few sheets of paper through the printer to clean the ink off the platen.

## THE MACHINE WILL NOT PRINT ANYTHING

- Be sure that the printer is plugged in and that no error light conditions are displayed.
- Be sure that the MultiPass server is loaded in.

■ Check print manager and delete all pending jobs, then retry the print operation.
■ If you are printing from a non-Windows program, be sure that the unit is online and in the Printer mode.

## YOU CANNOT PRINT FROM THE FILE MENU IN A WINDOWS APPLICATION

Check to be sure that the unit is correctly connected to the computer and that it is turned on. If it is still not printing, perform the following:

■ Open the control panel/printers in Windows.
■ Set the Canon MultiPass printer as the default printer.
■ Click the connect button to be sure that the correct port has been selected, then click OK.
■ Doubleclick the Canon MultiPass name. This sets the Canon MultiPass printer as the default printer.

## THE PRINTOUT IS TOO LIGHT

If you are printing in HS mode, the print quality might be too light or the print settings in Windows might be set to draft. Choose the HQ mode.

## DISCONNECTING THE PRINTER PORT

You might want to use the printer port on your computer for other equipment. If you want to do this, you must disable the MultiPass Server software before disconnecting the Multi-Pass printer.

To disconnect the MultiPass printer perform the following (7) items:

1 Turn off your computer.
2 Unplug your computer from all electrical sources.
3 Unplug the MultiPass printer from all electrical sources.
4 On the back of your computer, remove the cable connector from the parallel printer port.
5 On the MultiPass 1000, release the wire clips and remove the cable connectors from the port.
6 Now plug your computer back into the electrical outlet ac socket.
7 Then unplug the phone cable from the MultiPass 1000. Your Canon printer is now disconnected from your computer.

## UNINSTALLING THE MULTIPASS DESKTOP MANAGER

You must use the Uninstaller program when you want to remove the Desktop Manager and related scanner, printer, and fax drivers, and install a new program version or want to use another type printer with your computer.

## UNINSTALL PROGRAM FOR WINDOWS 95

Perform the following steps for uninstalling the MultiPass server for Windows 95.

1 Close the MultiPass Server.
2 Click the Start button and point to Programs. Click MultiPass Utilities. Then click MultiPass Uninstaller.
3 Then you follow the instructions that come up on your computer screen. When the program is completed, you can return to the desktop program. The files will all be deleted. If you want to install a new version, you can do so at this time.
4 To completely delete the MultiPass desktop manager, use the Windows Explorer.

## DIAGNOSING SOFTWARE AND HARDWARE PROBLEMS

If you are having some type of printer problem, you can use the MultiPass diagnostics to identify software configuration problems as well as hardware installation problems.

### Using MultiPass diagnostics for Windows 95

1 Click the Start button and point to Programs. Click MultiPass Utilities. Click MultiPass diagnostics. The program will then start and the diagnostics will begin. When the diagnostics are finished, a message appears, stating that all tests were performed successfully. If there were any problems, messages appear suggesting solutions.
2 A dialog box appears asking if you want to view a log file. The log file contains important trouble information and solutions.
3 Click Yes to view the log file. The MultiPass Diagnostics window appears.
4 To save this file, choose Save from the File menu. Select a drive and directory if you do not want the file saved in the MPASS directory. In the File Name box, type a name.
5 Click OK. The file is saved as a plain ASCII test file.
6 To exit this window, choose MultiPass Diagnostics from the Exit menu. The window closes and you return to the desktop.

# Plain-paper Fax-machine Operation

The plain-paper fax machine is used to transmit and receive images, printed or graphics, over regular telephone lines at a speed of approximately 10 seconds per page. A typical fax machine is shown in Fig. 10-14.

As you are given a simplified explanation of fax-machine operation, refer to Fig. 10-15. These blocks represent the scanner, printer, telephone modem, logic/microprocessor control unit, memory chips, and power supply.

The scanner is used to scan your document and send this digital data to the logic/microprocessor unit, where it is processed and sent to the modem. The modem is used to translate the digital signal into an audible sound or tones so that the fax information can be sent over a phone line. The reverse occurs when your fax machine receives a document. The modem receives the telephone tone signals and converts them into digital information that

**FIGURE 10-14**    **A typical fax machine.**

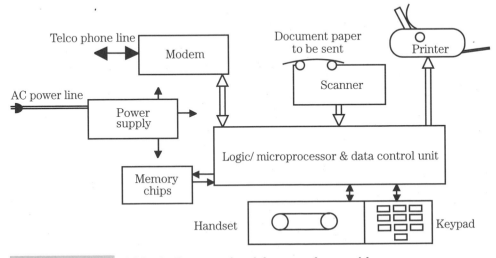

**FIGURE 10-15**    **A block diagram of a plain-paper fax machine.**

the logic/microprocessor unit can understand. The processed digital/logic information is then sent onto the printer, where it is then printed onto plain paper.

The memory chips are used as a buffer to store or hold data if it comes in over the phone line faster than it can be printed or if a document is scanned faster than the data can be sent out over the phone line. Also, it is used to store fax data if the printer runs out of paper, then it will print the fax when the paper cartridge holder is reloaded. The power section block is used to supply a regulated dc voltage to all of the other fax operational blocks.

## FAX MODEM OPERATION

Because digital information cannot be sent over your telephone line, a modem is used to convert the digital pulses into audible or tone signals. This continuous processing of modulation and demodulation between your fax machine and phone line is performed by a *modem (mo*dulation/*dem*odulation). The block diagram for a fax modem is shown in (Fig. 10-16).

## SOME FAX MODEM PROBLEMS

Look at some common fax problems that are actually modem- or phone-line related. You might have a problem where you cannot send or receive fax messages. This could be a very simple problem of the modem being "locked-up." The modem will actually stop the transfer of data. This lock-up will usually occur when you have had some power-line or phone-line glitches. This could be lightning-induced spikes or power surges. The modem can usually be unlocked by just removing the ac power to the fax machine for 1 or 2 minutes and then plugging it back in again. This will reset the modem operation again.

Noise on the phone line can cause modem lock-up, a brief interruption of the fax message, or garbled printed text. This same problem might occur if you have too many phone devices connected onto one phone line and are loading it down.

Some fax machines can be programmed to automatically switch over to the fax machine when any fax tones are received, then switch back to your phone when the fax is completed. This allows you to only need one phone line. You can also have a special "ring" programmed from your local phone company to let you know if you have a phone call or a fax coming to your phone line. This special ring will also cause the fax machine to come on line for a fax message. If your fax machine does not have these features and you only

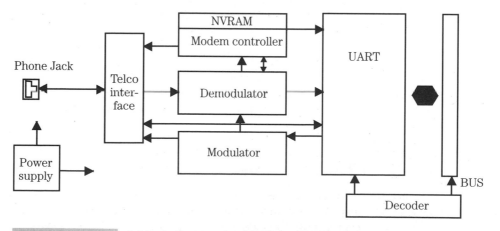

**FIGURE 10-16**    A block diagram for a fax machine modem.

**FIGURE 10-17**    **With the Radio Shack automatic fax transfer switch, you only need one phone line for voice phone and fax machine operation.**

have one phone line, you can install an automatic fax/phone switch. The fax switch (Fig. 10-17) can be purchased at Radio Shack and will automatically switch from the phone to the fax machine.

## FAX MACHINE OPERATIONAL PANEL

A typical fax-machine control panel is shown in Fig. 10-18. Refer to the numbered key points and see what these various controls are used for.

You will use the Start/Copy button to send a fax after dialing up the fax number. The Stop button is used to stop the machine for any reason, such as paper being jammed, empty ink cartridge, etc.

- *Printer button*  Use this button when you need to perform print head cleaning or when you want to print from a non-Windows application.
- *Print error light*  It lights when a paper jam occurs, or when sending or receiving a fax.
- *Printer lights*  These lights will indicate the status of the fax printer.
- *LCD display*  Displays messages, print errors, and other fax machine settings.
- *Function buttons and lights*  Use these buttons for fax and telephone operations. The lights indicate the status of the fax machine.
- *Speaker volume switch*  Use this switch to adjust the speaker's volume. This switch works in conjunction with the On-hook button.
- *Alarm light*  Flashes when an error occurs, when the printer is out of paper, out of ink, or when a received fax document is stored in memory.
- *One-touch speed dialing keypad*  Use these buttons for one-touch speed dialing and to perform special operations.
- *Printer panel cover*  Lift to access the printer panel, which you use to control fax printer operations.

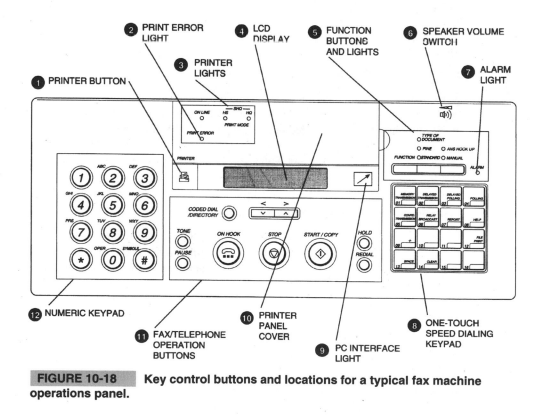

**FIGURE 10-18**    **Key control buttons and locations for a typical fax machine operations panel.**

- *Fax/telephone operation buttons*  Use these buttons for fax and telephone operations. On-hook or off-hook operations.
- *Numeric keypad*  Use these buttons to enter numbers and names when loading information and to dial fax/telephone numbers that have not been entered for automatic dialing.

## SOME FAX PROBLEMS AND SOLUTIONS

The following list contains some common fax problems and solutions:

### You cannot send documents

- Be sure that you are feeding the paper properly into the automatic document-feeder device.
- Check and see if the receiving fax machine has paper installed, machine is turned on and on line, and the fax machine is in the Receive mode.
- Check to hear a dial tone when you lift the handset.
- Be sure that the dialing method, Touch-Tone or pulse, is set correctly.

### The images you have sent are dirty or spotted

- Be sure that the document scanning glass is clean.
- Properly clean the scanning glass if it is dirty.

### You cannot receive documents automatically

- Be sure that all fax machine connections are tight and clean.
- Be sure that the fax machine is set to receive documents automatically.
- Be sure that you have printed out any documents that have been received and stored in memory.
- Be sure that paper is installed in the paper cassette holder.
- Check any of the read-out displays for any error messages, then clear them.

### You cannot receive documents manually

- Be sure that you have not fed a document into the automatic document feeder.
- Be sure that you press Start/Copy before hanging up the phone receiver.
- Be sure that you have printed out any documents in memory before sending or receiving manually.
- Check any of the read-out displays for any error messages and clear them out.

### Nothing appears on the printed page

- Clean the print head several times.
- Check and be sure that the ink cartridge is properly installed.
- Be sure that the Ink detector option is set to On.
- Still not printing? Then install a new ink cartridge.

### You cannot make copies

■ Be sure that the handset is on the hook.

■ Be sure that the document is set into the automatic document feeder.

■ If your fax machine has a self-diagnosis feature, then print out an activity report and see if any faults have occurred that need to be corrected.

### Fax machine will not work (dead)

■ The fax machine might have overheated and has shut itself down. Let the machine cool down and then try again.

■ Unplug the fax machine. Wait 20 to 30 seconds and then plug it back into the ac socket. Now try to operate it again.

■ At times, the problem might be with the party's fax machine you are sending to. If yours works with other machines, have the other party check out their fax machine. Also, be sure that you both have compatible fax machines.

**Fax machine paper jammed**    For the following jammed paper problems, refer to Fig. 10-19.

*Problem*    The paper document is jammed.
*Solution*    Remove the paper document you are trying to send and start again.

*Problem*    The fax machine tried to receive instead of send because you did not feed the paper in properly.
*Solution*    Feed the paper document into the machine and start the operation again.

*Problem*    Your fax machine tried to poll another unit, but the other fax machine did not have a document to send.
*Solution*    Contact the other party and have them set their fax machine document for polling.

*Problem*    The paper has become jammed.
*Solution*    Clear the paper jam (Fig. 10-19).

*Problem*    The paper cassette has not been completely inserted into the fax machine.
*Solution*    Insert the paper cassette all the way into the machine, then press Stop.

### Paper jammed in printer area

*Problem*    With the paper jammed in the printer area you should open the top cover. The printer cartridge will move over to the left side.
*Solution*    Now gently pull the jammed paper out the top. Be sure the printer light is off. If the printer cartridge is not all the way to the left side, gently move the cartridge to the left and then remove the paper.

**Paper jammed in bottom of fax printer**    If you have looked in both the paper cassette area and the printer area and have not been able to locate the jammed paper, perform the following steps:

■ Remove the paper cassette tray and any document supports.

■ Tilt up the front of the fax machine.

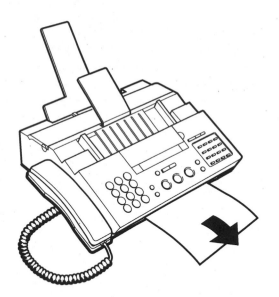

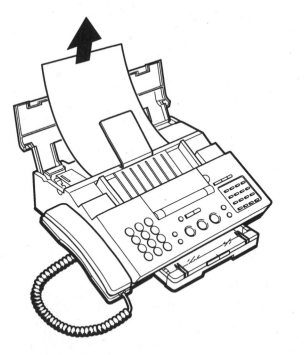

**FIGURE 10-19    Pull any jammed paper out of the machine in the direction shown by the arrows.**

- Now look into the opening at bottom of unit where the paper cassette was removed.
- If the paper is jammed back inside this opening, carefully remove it at this time.
- Now lower the Printer back on the table.
- Replace the paper cassette holder and any documents support items.
- Press the Stop button.

# Dot-matrix Printer Operation

The dot-matrix printer has been in use for many years for making hard copies from computers and home PCs. They are slower printing and quite noisy as the print head pins are fired, as compared to ink-jet and laser printers. The dot-matrix printer is lower in cost, uses an ink ribbon and can make multiplE paper copies. The dot-matrix printer produces characters by firing a bundle of plunger pins, between 9 and 24, with magnetic coils onto inked ribbon. These dots are all generated within the print head by electrical pulses from the printer's microprocessor.

## DOT-MATRIX PRINTER BLOCK DIAGRAM

The dot-matrix printer contains six main blocks (Fig. 10-20). These block's operations are as follows:

1 Microprocessor printer control unit.
2 The paper-transport device.
3 Printer ribbon transport.
4 The dot-matrix print-head assembly.
5 Carriage transport system.
6 The power supply.

## PRINT-HEAD OPERATION

Each dot of the dot-matrix print head is formed by a print-pin that is magnetically pushed down by a coil solenoid (Fig. 10-21). An electronic current pulse from the control-unit ICs "shoots" the pin out by the coil's magnetic force. When no pulse is present, this will cause a loss of the magnetic field and the return spring pulls the pin back into place for the next pulse to occur. The coils and print pins are very small and are offset from each other within a bundle. The moving pin strikes an ink ribbon and thus marks the paper. The pin coils are, at all times, receiving different timing pulses for the characters to be printed as the print

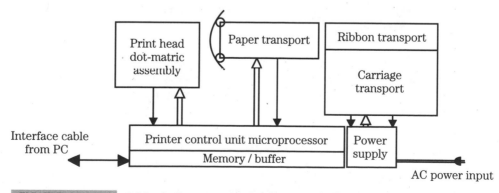

**FIGURE 10-20**   **A block diagram of a dot-matrix printer.**

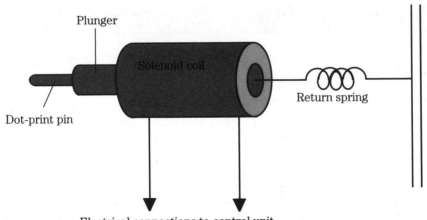

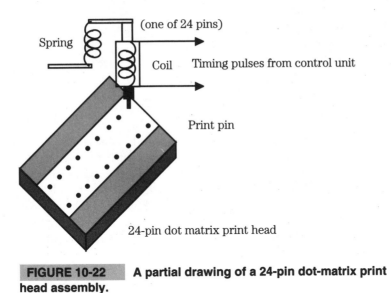

Plunger

Solenoid coil

Dot-print pin

Return spring

Electrical connections to control unit

**FIGURE 10-21**    **One print-head configuration of a 24-pin cluster mounted on a dot-matrix print-head assembly.**

(one of 24 pins)

Spring

Coil    Timing pulses from control unit

Print pin

24-pin dot matrix print head

**FIGURE 10-22**    **A partial drawing of a 24-pin dot-matrix print head assembly.**

head moves across the page. The results that proper characters are formed and printed on the paper. A partial drawing of a 24-pin dot-matrix print head assembly is illustrated in Fig. 10-22. The impact printing of the dot-matrix system is very noisy and the current required to drive all of the pin solenoid coils generates considerable heat.

## OVERALL SYSTEM OVERVIEW

Now take a brief look at the overall operation of your computer and dot-matrix printer operation. Refer to Fig. 10-23 to follow along on this simplified explanation.

When you type a page, the computer is putting digital codes into memory. When you "tell" your PC to print this information, it sends ASCII digital codes over the computer interface cable to the printer's buffer ICs. These ASCII codes gives the printer the proper characters to print, carriage returns, tabs, and other control information. Because your PC will "dump" the digital codes much faster than the printer can print, the buffer (which consists of RAM chips) is used to store this data into memory until it is called upon and can be used by the printer.

The printer's microprocessor takes the ASCII codes and properly processes them to activate the print-head pins, make carriage returns, control movement of the platen, and print head position. The current pulses generated from the processor actually activates (drives) each electromagnet pin within the print head to produce a readable text on the paper. To make a "Bold type format print," a second set of dots are offset slightly from the first ones.

To simplify, data is sent from your PC and is interpreted by the printer processor control unit and converted to a series of vertical dot patterns that is imprinted on the paper.

The operation of the friction paper-feed drive, the tractor paper-feed drive transport, the carriage transport, and ribbon transport systems are all very similar to the daisywheel and ink-jet printers that have been explained in the earlier part of this chapter. Please refer back to them for their operation and repairs.

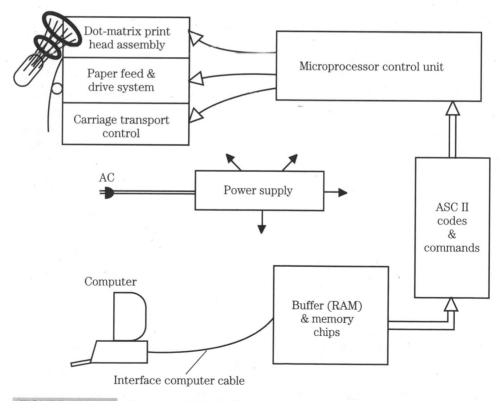

**FIGURE 10-23** The overall block diagram of a dot-matrix printer system.

### Printers/print head troubles and tips

*Problem*   Intermittent printing.
*Solution*   This could be caused by poor or intermittent print head cable connections. Unplug the printer and check and clean all cable connections and wires. Also, check any ground wire connections. Use an ohmmeter to check the cable wires for any open lead wires. Also, check the voltage from the power supply that goes to the print-head drivers.

*Problem*   Print head moves back and forth, but will not print.
*Solution*   Check the ribbon to be sure it is not out of ribbon and the cartridge is seated properly. Also, note if the ribbon is advancing properly or may have become jammed with the print-head pins. Clean, check, or install a new ribbon cartridge.

*Problem*   Printer will not print at all.
*Solution*   Be sure that the printer is plugged in, is receiving ac power, and check for any blown fuses in the power supply. If these items check out, then be sure that the printer is on line. An on-line light should be on the printer's control panel. The printer will not print if it is off line. The printer might also be out of paper. Next, recheck your computer printing programs. Then check the interface cable between your PC and the printer. Clean and tighten the plug-in and pin connections. Try another cable if you suspect it to be defective.

*Problem*   Print dots appear missing or faded.
*Solution*   The most common cause for poor or faded print quality is a ribbon problem. If in doubt, install a new ribbon cartridge. Also, the spacing on the print head might need to be adjusted. This adjustment can give your letters more intensity. Check and clean all parts of the pins and print head and be sure that all of the pins can be moved freely and do not stick.

*Problem*   Dots are missing. The missing dot syndrome might be an intermittent problem.
*Solution*   If the printer is working OK in all other respects, this problem is probably caused by one pin not firing, a stuck pin, current not getting to the firing coil, or a bent or broken pin. Inspect and clean the print head. Check for a stuck or bent pin. Check all wiring for any shorts or breaks. For a worst-case situation, you will have to replace the print-head assembly.

# How Laser Printers Work

The laser printer, also referred to as an *EP (ElectroPhotographic) printer* is not at all like the ink-jet or dot-matrix printers with a moving print-head carriage that were previously explained in this chapter. The laser printer uses a photosensitive drum, static electricity, pressure, heat and chemistry to produce sharp-printing house-quality detail copy. And the printing is fast and quiet.

To do this quality printing, the laser machine must perform these following operations, all at the same time:

■ Digital data language from your PC must be properly interpreted.
■ The PC data language and instructions are then processed to control the laser-beam modulation and movement.

- The paper movement must be precisely controlled.
- Drum position and rotation speed has to be controlled.
- Paper must be sensitized so that it will accept the toner that produces the printed images.
- The last step is to fuse the toner image to the paper with heat.
- The microprocessor dc controller must control all of these procedures flawlessly.

## LASER PRINTER BLOCK DIAGRAM OPERATION EXPLANATION

At the start of printing, the controller sends a command to load in a sheet of paper from the tray holder. As shown in Fig. 10-24, the paper is drawn through the paper pickup and feeder rollers. Then, as the paper goes around the drum, the laser beam starts to form the static image on the drum surface. (More details on the photo-sensitive drum later.)

The scanning mirror is used to deflect the laser beam back and forth across the surface of the cylinder drum. Also, during scanning, the beam is turned off and on from the modulation, which is controlled by whatever digital information is sent from your PC to the printer's microprocessor controller. All of this laser-beam action is what produces the proper dot pattern on the drum.

During this operation, toner (black powder) is applied to the drum. The toner has a negative electrical charge and is attracted to the positive-charged dots that the laser beam created on the drum's surface. The toner sticks because of an electrostatic charge on the drum in small dots to produce the printed pattern.

The paper, now with toner on it, receives the image from the drum and is then fed into the fuser system.

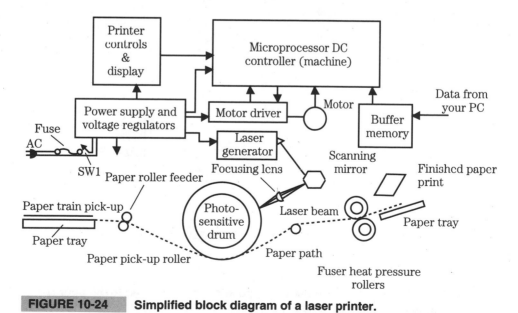

**FIGURE 10-24**   Simplified block diagram of a laser printer.

The motor drive and motor is used to turn the photosensitive drum. Also, a series of gears or belts from the motor turns the paper-pickup loader, feeder rollers, and the heated fuser rollers.

The information that you want printed from your PC goes to a buffer and then into the laser printer microprocessor or dc controller. The dc controller is the heart of the laser printer. The microprocessor controls all printer operations, such as paper travel, paper-tray amount, temperature of the fuser heating system, drum and paper speed, and the laser generator. The microprocessor controls the sweep of the laser beam and also quickly turns the laser light beam off and on to paint the image to be printed on the drum's surface. If problems develop the microprocessor will shut the printer down and, in some cases, give you an error on the control display panel.

The power supply develops all of the required voltages, some regulated dc voltages, to perform all of the printer operations. Power is required for drum rotation, paper-feeder rollers, pressure rollers, laser generator, scanning mirror, paper movement, and to heat the fuser roller.

## PHOTOSENSITIVE DRUM OPERATION AND CARE

Figure 10-25 illustrates in more detail of how the photosensitive drum produces images on paper inside the laser printer. Each time that a new sheet of paper is printed the drum must be electrically erased and cleaned of any toner particles. A rubber cleaning blade is used to remove any toner from previous images. An erase light is also used to neutralize the drum before the laser beam can produce a new image. In order for a new image to be written, the

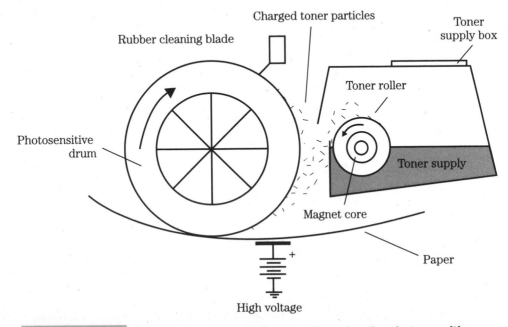

**FIGURE 10-25**   How the toner (powder) is transferred to the photosensitive drum and the paper image is developed in a laser printer.

drum must be charged or conditioned. A high-voltage static charge is used to condition the drum.

To rewrite this clean drum, a laser beam is swept back-and-forth across the drum's surface while the beam is turned on-and-off to create an image with a series of dots. This dot process is somewhat like the way a dot-matrix printer makes an image.

The rotating drum then passes very close to the toner roller. The toner roller, which has a magnetic core, has toner particles attracted to it, which, in turn, deposits the toner onto the drum's surface. The toner powder then sticks to the charged areas of the drum, which then creates the image to be printed.

Now the toner image on the drum surface must be developed. To do this, the toner is transferred from the drum surface onto the paper, which is fed under the drum. The drum rotation is held to the same travel speed as the paper. The toner is transferred to the paper with electrostatic ionized air, which acts as a magnet to the plastic resin and iron toner particles. The drum is then cleaned for the next image with the rubber cleaning blade.

Next, the toner-covered paper goes onto the fuser assembly to be permanently bonded to the paper. Heat and pressure is used for the fusing process. The paper passes between two rollers, the top roller is heated and melts the toner, while pressure from the bottom roller presses the toner into the paper fibers and the printed page is finished. The paper will feel warm after it is printed.

## LASER PRINTER PROBLEMS AND TIPS

The laser printer is a very sophisticated machine and all sections must perform correctly for you to have a crisp, sharply detailed print copy.

**Printer will not turn on (dead)**  First, check for the presence of ac power at printer socket. Is the laser printer plugged in? If this is OK, check for any blown fuses. If you have power and the cooling fans are working, check for any error readout messages for the self-test check. These error messages give you some clues of the problem. Some printer shutdown problems can be caused by a power-line "glitch," which can lock up the microprocessor and can be corrected by unplugging the printer power for 30 seconds and plugging it back in. This will reset the printer microprocessor controller.

The error message might indicate a communications problem between your PC and the laser printer. This might be caused by an incorrect software program or a fault in the data cable between the PC and printer. Check the cable, clean and tighten the plug pins and contacts. You might have to replace the cable with a new one.

**Paper is jammed or has tears**  Paper jammed or feeder problems could be caused by paper not placed in the tray properly or the wrong kind of paper being used. Other paper jam problems can be caused by loose belts, broken or worn gears, and misadjusted feed and pressure rollers. And do not overlook a defective toner cartridge for not only poor prints, but also paper jams.

**Prints have splashes and specks**  Turn the printer off. Check and clean the fusing rollers. Also, be sure that the fusing roller is being heated. The laser printer should be cleaned often because of the toner dust that is always present. Check the adjustment of the

**FIGURE 10-26** A laser copier with the cover tilted up so as to be cleaned or serviced.

**FIGURE 10-27** A laser printer that is used with a PC for printouts. The cover has been lifted up to clean and service the unit.

rubber cleaning blade and see if the cylinder drum is being cleaned properly because this will cause the copies not to be clean and sharp.

A laser copier is shown in Fig. 10-26 with the cover tilted up to clean and service the machine. The laser printer in Fig. 10-27 is used in conjunction with a PC to make hard-copy printouts. The cover has been lifted up to clean and service the unit.

As you can see, the laser printer is a very sophisticated electronic and mechanical printing machine. Except for minor repairs and cleaning, you might want to consult a qualified laser printer service technician or company. In many cases, some very expensive diagnostic equipment is needed to solve the laser printer problem. Use caution and work with care on any of the printers covered in this chapter.

# GLOSSARY

Use this glossary to help you better understand some of the terms used in explaining "How Electronic Things Work" in these book chapters.

# VCR, Camcorder, Audio Tape

This glossary section can be used in conjunction with the video recorder, camcorder, and audio tape recorder chapters.

*acoustic suspension*  Air-suspension (AS) speakers are sealed in an enclosure or box to produce natural, low-distortion base output. Greater driving power is needed with these less-efficient speaker systems.

*air suspension*  Another name for an acoustic-suspension speaker.

*amp*  Abbreviation for amplifier.

*ANRS*  A noise-reduction system that operates on principles that are similar to the Dolby system.

*APC*  The automatic power control circuit keeps the laser-diode optical output at a constant level in the CD player.

*audio/video control center*  The central control system that controls all audio and VCR operations.

*auto eject*  The tape player feature that automatically ejects the cassette at the end of the playing time.

*auto focus*  AF is the focus servo that moves the objective lens up or down to correct the focus of the CD player.

*auto record level*  Automatic control of the recording level.

*auto reverse*   The ability of the cassette player to automatically reverse directions to play other side of the tape.

*auto tape selector*   Automatic bias and equalization when the cassette is inserted into the tape.

*azimuth*   The angle at which the tape head meets the moving tape. A loss of high-frequency response is often caused by an improper azimuth adjustment.

*azimuth control*   A control to adjust the angle of the tape control to correct misalignment in the auto stereo tape player.

*baffle*   The board on which the speakers are mounted.

*balance*   The control in the stereo amp that equalizes the output audio in each channel.

*bass reflex*   A bass-reflex system vents backward sound waves through a tuned vent or port to improve bass response.

*bias*   A high-frequency current applied to the tape-head winding to prevent low distortion and noise while recording.

*block diagram*   A diagram that shows the different stages of a system.

*booster amplifier*   A separate amplifier that is connected between the main unit and the speakers in a car stereo system.

bridging   Combining both stereo channels of the amp to produce a mono signal with almost twice the normal power rating in a car stereo system.

*cabinet*   A box that contains speakers or electronic equipment.

*capstan*   The shaft that rotates against the tape at a constant rate of speed and moves the tape past the tape heads. In the cassette player, a rubber pinch roller holds the tape against the capstan.

*cassette radio*   The combination of an AM/FM tuner, amplifier, and cassette player in one unit.

*cassette tuner*   A tuner and cassette deck in one chassis.

*CH*   The abbreviation for channel. The stereo component has two channels (left and right).

*channel separation*   The degree of isolation between the left and right channels, often impressed in decibels. The higher the decibel values, the better the separation.

*chassis*   The framework that holds the working parts in the amplifier, tuner, radio, cassette, CD player, or VCR. The chassis could be metal, plastic, or a PC board.

*chips*   Chip devices can contain resistors, multilayer ceramic chip capacitors, mini-mold chip transistors, mini-mold chip diodes, and mini-mold chip ICs.

*clipping*   Removing or cutting off the signal from a waveform that contains distortion, which can be seen on the oscilloscope. Excessive power results in distortion.

*coaxial speaker*   A speaker with two drivers mounted on the same frame. The tweeter is mounted in front of the woofer speaker. Usually, coaxial speakers are used in the car audio system.

*compact disc*   The compact-disc (CD) player plays a small disc of digitally encoded music. The CD provides noiseless high-fidelity music on one side of a rainbow-like surface.

*CPU*   A computer-type processor used in the master and control mechanism circuits of a CD player.

*crossover*   A filter that divides the signal to the speaker into two or more frequency ranges. The high frequencies go to the tweeter and the low frequencies go to the woofer.

*crosstalk*   Leakage of one channel into the other. Improper adjustment of the head might cause crosstalk between two different tracks.

*D/A converter*   In the CD player, the device that converts the digital signal to an analog or audio signal.

*dc*   Direct current is found in automobile battery systems, and also after the ac has been filtered and rectified in low-voltage power supplies.

*decibel*   The decibel (dB) is a measure of gain, the ratio of the output power or voltage, with respect to the input (expressed in log-units).

*de-emphasis*   A form of equalization in FM tuners to improve the overall signal-to-noise ratio while maintaining the uniform frequency response. The de-emphasis stage follows the D/A converter in a CD player.

*dew*   A warning light that might come on in a VCR or camcorder. It indicates too much moisture at the tape head.

*digital*   Within tuners, the digital system is a very precise way to lock in a station without drifting. Digital recording is used in compact discs.

direct drive   A direct-drive motor shaft is connected to a spindle or capstan/fly wheel. The CD rests directly on the disc or spindle motor in CD players.

disc holder   The disc holder or turntable sits directly on top of the motor shaft in the CD player.

*dispersion*   1. The spread of speaker high frequencies, measured in degrees. 2. The angle by which the speaker radiates its sound.

*distortion*   In a simple sine-wave signal, distortion appears as multiples (harmonics of the input frequency). A type of distortion is the clipping of the audio signal in the audio amplifier.

*Dolby noise reduction*   A type of noise reduction that works by increasing the treble sounds during recording and decreasing them during playback, thus restoring the signal to the original level and eliminating tape hiss.

*driver*   1. In a speaker system, each separate speaker is sometimes called a *driver*. 2. The loading, feed, and disc motors might be driven by transistor or IC drivers.

*drive system*   The motors, belts, and gears that drive the capstan/flywheel in cassette tape or CD players.

*dropout*   In tape decks, dropouts occur when the tape does not contact the tape head for an instant. Dropouts occur in the compact disc because of dust, dirt, or deep scratches on the plastic disc.

*dual capstan*   Dual capstans and flywheels are used in auto-reverse cassette players and can play tapes in both directions.

*dynamic*   A dynamic speaker has a voice coil that carries the signal current with a fixed magnetic field (PM magnet), and moves the coil and cone. The same principle applies to the human ear or to headphones.

*dynamic range*   The ratio between the maximum signal-level range and the minimum level, expressed in decibels (dB).

*electronic speed control*   An electronic method of controlling the speed of the capstan motor.

*electrostatic*   An electrostatic speaker headphone, or microphone, that uses a thin diaphragm with a voltage applied to it. The electrostatic field is varied by the voltage, which moves the diaphragm to create sound.

*equalizer*   A device to change the volume of certain frequencies, in relation to the rest of the frequency range. Sliding controls can be found in auto-radio and cassette-player equalizers.

*erase head*   A magnetic component with applied voltage or current to remove the previous recording or noises on the tape. The erase head is mounted ahead of the regular R/P head.

*extended play*   EP refers to the six hours of playing time that is obtainable with a T-120 VHS cassette played in a VCR.

*eye pattern*   The RF signal waveform at the RF amplifier in a CD player. The waveform is adjusted to a clear and distinct diamond-shaped pattern.

*fader*   A control in auto radio or cassette players to control the volume balance between the front and rear speakers.

*fast forward*   The motor in the cassette, VCR, or CD player can rotate faster with a higher voltage applied to the motor terminals or when larger idler pulleys are pushed into operation.

*filter*   A circuit that selectively attenuates certain frequencies, but not others. The large electrolytic capacitor in the low-voltage power supply is sometimes called a *filter capacitor*.

*flutter*   A change in the speed of a tape transport, also known as *wow*.

*focus error*   The output from the four optosensing elements are supplied to the error signal amplifier and a zero output is produced. The error amp corrects the signal voltage and sends to the servo IC to correct the focus in the CD player.

*folded horn speaker*   The system that efficiently forces the sound of the driver to take a different path to the listener.

*frequency response*   The range of frequencies that a given piece of equipment can pass to the listener. The frequency response of an amplifier might be 20 Hz to 20 kHz.

*gain*   The amplification of an electronic signal. Gain is given in decibels.

*gain control*   A control to adjust the amount or boost the amount of signal.

*gap*   The crucial distance between the pole pieces of the tape head. The gap area might be full of oxide, which would cause weak, distorted, or noisy reception.

*glitch*   A form of audio or video noise or distortion that suddenly appears and disappears during VCR operation.

*graphic equalizer*   An equalizer with a series of sliders that provides a visual graphic display.

*ground*   A point of zero voltage within the circuit. The common ground might be a metal chassis in the amplifier. American-made cars have a negative-ground polarity.

*head*   A magnetized component with a gap area that picks up signals from the revolving tape.

*hertz*   Hertz (Hz) are the number of cycles per second (CPS), the unit of frequency.

*hiss*   The annoying high-frequency background noise in tapes and record players.

*hum*   A type of noise that originates from power lines, caused mainly by poor filtering in the low-voltage power supply. Hum and vibrating noise might be heard in transformers or motors that have loose particles or laminations.

*idler*   A wheel found in tape players to determine the speed of the capstan/flywheel or turntables in the cassette player.

*impedance*   The degree of resistance (in ohms), that an electrical current will encounter in a given circuit or component. A speaker might have an impedance of 2, 4, 8, 16, or 32 ohms.

*integrated circuit*   An IC is a single component that has many parts. ICs are used throughout most cassette players, amplifiers, VCRs, and CD players.

*interlock*  A safety interlock device used in the CD player to load the disc.

*ips*  Inches per second, the measurement of cassette-tape speed.

*jack*  The female part of a plug and receptacle.

*kilohertz*  1 kHz is equal to 1000 Hz.

*laser assembly*  The assembly that contains the laser diodes, focus, and tracking coils in a CD player.

*laser current*  Low laser current might indicate that a laser-diode assembly in the CD player is defective.

*laser diodes*  The diodes that pick up the coded information from the disc along with the optical pick-up assembly in a CD player.

*LED*  Light-emitting diodes are used for optical readouts and displays in electronic equipment.

*level*  1. The strength of a signal. 2. The alignment of the tape head with the tape.

*line*  Line output or input jacks are used in the amplifier, cassette, or CD player. The line signal is usually a high-level signal.

*loading motor*  The motor in CD players, VCRs, and camcorders that moves the tray or lid out and in so that the disc or cassette can be loaded.

*long play*  LP is a speed on the VCR that provides four hours of recording on a 120-minute VHS cassette.

*loudness*  The volume of sound. Loudness is controlled by a volume control.

*LSI*  Large-scale integrated circuits include processors, ICs, and CPUs that are used in VCRs, camcorders, and disc players.

*magnetic*  Metal attraction. The magnetic coil might be found in the VOM or VTVM.

*megahertz*  1 MHz is equal to 1000 kHz or 1,000,000 Hz.

*memory*  The program memory of a CD player.

*metal tape*  The high-frequency response and maximum-output level are greatly improved with metal tape. Pure metal cassettes are more expensive than the regular oxide cassettes.

*microprocessor*  A multifunction chip found in most of today's electronic products. They are used in tape decks, transports, memory operations, CD players, and VCRs.

*monitor*  To compare signals. A stereo amplifier can be monitored to compare the signal with the defective channel.

*monophonic*  One channel of audio, such as in a single speaker.

*multiplex*  A multiplex (MPX) demodulator in the FM tuner or receiver converts a single-carrier signal into two stereo channels of audio.

*mute switch*  The mute switch might be a transistor in the audio-output line circuit of a CD player or cassette deck.

*noise*  Any unwanted signal that is related to the desired signal. Noise can be generated during the record and play functions in a cassette player. A defective transistor or IC could cause a frying noise in the audio.

*NR*  Noise reduction.

*optical lens*  The lens located in the pick-up head of a CD player. Clean the lens with solution and a photographic dry-cleaning brush.

*output power*  The output power of an amplifier, rated in watts.

*oxide*  The magnetic coating compound of the recording tape or cassette. The excess oxide should be cleaned off of the tape heads, pinch rollers, and capstans for good music reproduction.

*passive radiator*   A second woofer cone that is added without a voice coil in the speaker cabinet. The pressure created by the second cone produces heavy bass tones.

*pause control*   A feature to stop the tape movement without switching the machine. The pause control is used in cassette decks, VCRs, and CD players.

*peak*   The level of power or signal. A peak indicator light shows that the signal levels are exceeding the recorder's ability to handle the peaks without distorting.

*phase*   Sound waves are in sync with one another. Speakers should be wired in phase.

*pick-up motor*   The pick-up, SLED, or feed motor is used to move the pick-up assembly in the radial direction or toward the outer edge of the disc.

*pitch control*   A control that changes the speed of the control motor.

*PLL*   The phase-locked loop (PLL) VCO circuit is used in the digital-control processor of the CD player with a crystal.

*port*   An opening in a speaker enclosure or cabinet. The port permits the back bass radiation to be combined with the front radiation for total response.

*power*   The output power of any amp is given in watts. A low-voltage power supply provides voltage to other circuits.

*preamplifier*   The amp within the cassette player that takes the weak signal from the tape head and amplifies it for the AF stages.

*rated power bandwidth*   The frequency range over which the amplifier supplies a certain minimum power factor, usually from 20 to 20,000 Hz.

*recording-level meter*   The meter (analog, LED, or fluorescent panel) that indicates how much signal is being recorded on the tape.

*reject lever*   A lever that rejects or deletes a given track in a cassette or a record on the record changer.

*remote control*   A means to operate the receiver, CD player, cassette/tuner, or VCR from a distance. Today, most remote-controlled transmitters are infrared type.

*repeat button*   The button that replays the same track of music on the CD player.

*RF*   A radio-frequency signal.

*ribbon speaker*   A high-frequency driver or tweeter speaker that uses a ribbon material suspended in a magnetic field to generate sound current when current is passed through it.

*saturation*   Recording tape is saturated when it cannot hold anymore magnetic information.

*self erase*   A degrading or partial erasure of information on magnetic tape.

*self-powered speakers*   A speaker with a built-in amplifier.

*separation*   The separation of two stereo channels. Placement of the stereo speakers can provide good or poor stereo separation.

*servo*   The tracking circuits that keep the laser pickup in the grooves at all times.

*servo control*   The servo control IC that controls the focus and tracking coils in CD players.

*signal processing*   In the CD player, converting the processing laser signals to audio with preamps and signal processors.

*signal-to-noise ratio*   The ratio (S/N) of the loudest signal to noise. The higher the signal-to-noise ratio, the better the sound.

*skewing*   A form of visual distortion or bend at the upper part of the picture of the VCR player.

*solenoid*   A switch that consists of an electric coil with an iron-core plunger that is pulled inside the coil by the magnetic field. Solenoids are usually found in auto radios, cassette, tape, and CD players.

*speaker enclosure*   The cabinet in which speakers are mounted.

*spindle motor*   The disc or turntable motor revolves.

*standard play*   SP is the speed at which a two-hour (T-120) VHS cassette plays on VCR machine.

*subwoofer*   A speaker that is designed to handle very low frequencies below 150 Hz.

*test cassette*   The recorded signals on a test cassette that are used for alignment and adjustment procedures on the cassette player.

*test disc*   A CD that is used to make alignments and adjustments in CD players.

*tone control*   A circuit that is designed to increase or decrease the amplification in a specific frequency range.

*tracking servo*   The IC that keeps the laser beam in focus and tracking correctly.

*tray*   The loading tray in which the CD to be played is placed.

*tweeter*   A high-frequency driver speaker.

*VCR*   Video cassette recorder.

*vented speaker system*   Any speaker cabinet with a hole or port to let the back waves of the woofer speaker escape. A bass reflex is a type of vented speaker system.

*VHS*   The system used today by most VCRs.

*voice coil*   The coil of wire that is wound over the end of the cone of the speaker in which the amplifier output is connected. The electrical signal is converted to mechanical energy to create audible sound waves.

*watts*   The practical unit of electricity and other power.

*woofer*   The largest speaker in a speaker system. The one that reproduces the low frequencies.

*wow*   A slow-speed fluctuation in tape speed. Fast-speed variation is called *flutter*.

# Telephone and Answering Machines

This glossary section can be used in conjunction with the telephone and answering machine chapter.

*ADC (analog-to-digital converter)*   An electronic device used to convert an analog voltage into a corresponding digital representation.

*AF (audio frequencies)*   The frequencies that fall within the range of human hearing, typically 50 to 18,000 Hz.

*AM (amplitude modulation)*   A technique of modulating a carrier sinusoid with information for transmission.

*anode*   The positive electrode of a two-terminal electronic device.

*attenuation*   The loss of reduction in a signal's strength because of intentional or unintentional conditions.

*bandwidth*   The range of frequencies over which a circuit or system is capable of operating or is allowed to operate.

*base*   One of three electrodes of a bipolar transistor.

*battery*   The operating voltage supplied to a telephone from a central office.

*BOC (Bell Operating Company)*   The local telephone company that provides your telephone service from your central office.

*capacitance*  The measure of a device's ability to store an electric charge, measured in farads, microfarads, and picofarads.

*capacitor*  A device used to store an electric charge.

*cathode*  The negative electrode of a two-terminal electronic device.

*cell*  In cellular telephony, the geographic area served by one transmitter/receiver station.

*channel*  An electronic communication path. A channel can consist of fixed wiring or a radio link. A channel has some bandwidth, depending on the type and purpose of the channel.

*CO (Central Office)*  The building and electronic equipment owned and operated by your local telephone company that provides service to your telephone.

*collector*  One of three electrodes on a bipolar transistor.

*continuity*  The integrity of a connection measured as a very low (ideally zero) resistance by an ohmmeter.

*CPC (Calling Party Control)*  A brief dc signal generated by your local central office when a caller hangs up.

*CPU (central processing unit)*  Also called a *microprocessor*. A complex programmable logic device that performs various logical operations and calculations based on predetermined program instructions.

*cradle*  An area on a telephone's housing where the handset or portable unit can be kept when not in use.

*DAC (digital-to-analog converter)*  An electronic device used to convert a pattern of digital information into a corresponding analog voltage.

*data*  In telephone systems, any information other than human speech.

*decibel (dB)*  A unit of relative power or voltage expressed as a logarithmic ratio of two values.

*demarcation point*  The point where a building connects with the outside wiring supported by the BOC. In a home, the demarcation point would be at the network interface connector.

*demodulation*  The process of extracting useful information or speech from a modulated carrier signal.

*diode*  A two-terminal electronic device used to conduct current in one direction only.

*drain*  One of three electrodes on a MOS transistor.

*DTMF (Dual-Tone Multi-Frequency)*  A process of dialing that uses unique sets of audible tones to represent the desired digit.

*emitter*  One of three electrodes on a bipolar transistor.

*EPROM (Electrically Programmable Read-Only Memory)*  An advanced type of ROM that can be erased and reused many times.

*Exchange area*  A territory in which telephone service is provided without extra charge. Also called the *local calling area*.

*FM (Frequency Modulation)*  A technique of modulating a carrier sinusoid with information for transmission.

*full-duplex*  A circuit that carries information in both directions simultaneously.

*gate*  One of three electrodes on a MOS transistor.

*ground start*  A method of signaling between a telephone and the central office, where a signal line is grounded to request service.

*half duplex*  A circuit that carries information in both directions, but in only one direction at a time.

*harmonics*   Multiples of some intended frequency, usually created unintentionally when a frequency is first generated.

*hybrid*   Also known as an *induction coil*. A specialized type of transformer used in classic telephones to couple the two-wire telephone line to an individual transmitter and receiver.

*ICM (incoming message)*   The message that is left by a caller on an answering machine.

*IF (intermediate frequency)*   A high-frequency signal used in the process of RF demodulation.

*impedance*   A measure of a circuit's resistance to an ac signal, usually measured in ohms or kilohms.

*inductance*   The measure of a device's ability to store a magnetic charge, measured in henries, millihenries, or microhenries.

*inductor*   A device used to store a magnetic charge.

*LCD (liquid-crystal display)*   A type of display using electric fields to excite areas of liquid crystal material.

*LED (light-emitting diode)*   A specialized type of diode that emits light when current is passed through it in the proper direction.

*loop current*   The amount of current flowing in the local loop.

*loop start*   The typical method of signaling an off-hook or line-seizure condition where current flow in the loop indicates a request for service.

*local loop*   The complete wiring circuit from a central office to an individual telephone.

*modulation*   The systematic changing of the characteristics of an electronic signal in which a second signal is used to convey useful information.

*MTS (Message Telephone Service)*   The official name for long-distance or toll service.

*NAM (Number Assignment Module)*   An erasable memory IC programmed with an assigned telephone number and specific identification information, typically used with cellular telephone circuits.

*OGM (Outgoing Message)*   The message that a caller hears when an answering machine picks up the telephone line.

*permeable*   The ability of a material to become magnetized.

*piezoelectric*   The property of certain materials to vibrate when voltage is applied to them.

*pps (Pulses Per Second)*   The rate at which rotary or pulse interruptions are generated. A rate of 10 pps is typical.

*program*   A sequence of fixed instructions used to operate a CPU.

*PSTN (Public Switched Telephone Network)*   A general term for the standard telephone network in the United States. The term refers to all types of wiring and facilities.

*pulse*   A process of dialing using an IC (instead of a mechanical device) to generate circuit interruptions corresponding to the desired digits.

*RAM (random-access memory)*   A temporary memory device used to store digital information.

*RC (Regional Center)*   Telephone facilities that interconnect both toll centers and some central offices, and support long-distance telephone service.

*rectification*   The process of converting dual-polarity signals to a single polarity.

*regulator*   An electronic device used to control the output voltage or current of a circuit, usually of a power supply.

*resistance*   The measure of a device's ability to limit electrical current, measured in ohms, kilohms, or megaohms.

*resistor*  A device used to limit the flow of electrical current.

*ring*  An alerting signal sent from a central office to a telephone or other receiving equipment, such as an answering machine.

*RF (radio frequency)*  A broad category of frequencies in the range above human hearing, but below the spectrum of light, typically from 100 kHz to more than 1 GHz.

*ring*  One of the two main wires of a local loop. The name originally referred to the ring portion of a phono plug that operators used to complete connections manually. See tip below.

*ROM (read-only memory)*  A permanent memory device used to store digital information.

*rotary*  A process of dialing that uses a mechanical device to open and close a set of contacts in a pattern corresponding to a desired digit.

*sidetone*  A small portion of transmitted speech that is passed to the receiver. It allows a speaker to hear their own voice and gauge how loudly to speak.

*SMT (surface-mount technology)*  The technique of PC board fabrication using components that are mounted directly to the surface of a PC board instead of inserting them through holes in the board.

*SOT (small-outline transistor)*  A transistor designed for use with surface-mount PC boards.

*source*  One of three electrodes on a MOS transistor.

*subscriber loop*  Another term for the local TC (toll center) facilities that interconnect central offices.

*tip*  One of the two main wires in a local loop. The name originally referred to the tip of a phono plug that operators used to complete connections manually.

*transistor*  A three-terminal electronic device whose output signal is proportional to its input signal. A transistor can act as an amplifier or a switch.

*transformer*  A device using inductors to alter ac voltage and ac current levels or to isolate one ac circuit from another.

*VOX (voice-operated control actuation)*  A circuit that detects the presence of a caller's voice and allows the machine to continue recording.

# Color TVs and Monitors

This glossary section can be used in conjunction with the color TV and monitor chapter.

*ac (alternating current)*  The type of electricity normally used in homes and most industries. Its contrasting opposite is direct current (dc), now obsolete except for certain specialized applications. All batteries supply dc.

*ACC (automatic color control)*  A circuit similar in function and purpose to AGC, except that it is supplied exclusively to the color bandpass amplifiers to maintain constant signals.

*ac hum*  A low-pitch sound heard whenever ac power is converted into sound, intentionally or accidentally. The common ac hum is 60 Hz.

*AFC (automatic frequency control)*  A method of maintaining the frequency or timing of an electrical signal in precise agreement with some standard. In FM receivers, AFC keeps the receiver tuned exactly to the desired station. In TV, horizontal AFC keeps the individual elements or particles of the picture information in precise register with the picture transmitted by the TV station.

*AGC (automatic gain control)*  A system that automatically holds the level or strength of a signal (picture or sound) at a predetermined level, compensating for variations caused by fading, etc.

*amplifier*  As applied to electronics, a magnifier. A simple tube or transistor or a complete assembly of tubes or transistors and other components can function as an amplifier of either electric voltage or current.

*antenna*  A self-contained dipole or outside device to collect the broadcast signal from the TV station. The collected signal is fed to the TV with a shielded or unshielded lead-in wire.

*anode*  The positive (+) element of a two-element device, such as a vacuum tube or a semiconductor diode. In a television tube, an anode is an element having a relatively high positive voltage applied to it.

*aperture mask*  An opaque disk behind the faceplate of a color picture tube; it has a precise pattern of holes, through which the electron beams are directed to the color dots on the screen.

*arc*  An electric spark that jumps (usually due to a defect) between two points in a circuit that are supposed to be insulated from each other, but not adequately so.

*aspect ratio*  The relation or proportion between the width and height of a transmitted TV scene. The standard aspect ratio is 4:3, meaning that the picture is three inches high for every four inches of width (four-thirds as wide as it is high).

*audio*  Any sound (mechanical) or sound frequency (electrical) that is capable of being heard is considered as audio. Generally, this includes frequencies between about 20 and 20,000 Hz.

*b+*  Supply voltage, as low as 1 Vdc in transistorized circuits and as high as hundreds of volts in tube circuits, which is essential to normal operation of these devices. The plus sign indicates the polarity.

*B+ boost*  A circuit in TVs, which adds to, or boosts, the basic B+ voltage. The boost source is a by product of the horizontal deflection system. Also see *damper*.

*bandpass amplifier*  In a color TV, one or two color signal amplifiers located at the beginning of the color portion of the TV; they are designed to amplify only the required color frequencies. They pass a certain band of frequencies.

*blanking*  A term used to describe the process that prevents certain lines and symbols (required for keeping the picture in step with the transmitter), from being seen on the TV screen.

*bridge rectifier*  Four diodes are wired in a series circuit to provide full wave rectification of a two lead power transformer. The ac-dc TV chassis may use a bridge rectifier after the line fuse.

*brightness*  Refers to both the amount of illumination on the screen (other than picture strength) and the control that is used to adjust the brightness level.

*burst*  In color TV, a precise timing signal. It is not continuous, but comes in spaced bursts. It is transmitted for controlling the 3.58 MHz oscillator essential for color reception.

*burst oscillator*  The precision 3.58 MHz oscillator vital to color reception. It is kept in step (sync) by the burst.

*buzz*  This is sometimes called intercarrier buzz, a raspy version of ac hum, usually caused by improper adjustment of some IF circuits.

*B-Y*  The blue component of a color picture minus the monochrome.

*capacitance*  A measure of a capacitor's ability to store electrical energy. The capacitor was called a condenser at one time. Bypass and electrolytic filter capacitors are found in many TV circuits. The unit of capacitance is the farad.

*carrier*  The radio signal that carries the sound or picture information from the transmitter to the receiver. The carrier frequency is the identifying frequency of the station (e.g., 880 kHz, 93.1 MHz, etc.)

*cathode-ray tube*  A tube in which electrical energy is converted to light. An electron beam (or beams), originating at the cathode, impinges upon a phosphor light-emitting screen. TV picture tubes, radar tubes, tuning eyes in some FM sets, and many similar types are basically cathode-ray tubes.

*chassis*  The base where the majority of electronic components are mounted. The metal chassis might be common ground. Today, in the solid-state chassis, the PC board wiring is the main chassis.

*cheater cord*  An ac line cord for operating the TV without the back cover or the cabinet when troubleshooting and repairing. The original cord is attached to the back of the cabinet as a safety measure.

*chroma*  Another term for color. Color amplifiers are often called *chroma amplifiers*. The term is also used to denote the control used to increase or reduce the color content of a picture.

*chopper circuit*  The chopper power supply is a pulse-width-modulated (PWM), regulated power supply. The chopper supply circuits are quite similar to the horizontal deflection system.

*circuit breaker*  The circuit breaker might work in place of the line fuse to open when an overload is in the TV circuits. Some horizontal output tubes have a separate circuit breaker in the cathode circuit.

*clipper*  A term describing the operation of one of the sync circuits in a TV. It is the stage (tube or transistor) that separates the sync (timing) signals from the picture information.

*color bar generator*  The color bar generator provides patterns for color alignment and color TV adjustments. Some of the NTSC generators have from 8 to 10 different patterns.

*color killer*  A special circuit whose function is to turn off the color amplifier circuits when a black-and-white signal is being received. This is also the name of the control used to adjust the operation of the circuit.

*comb filter*  The comb filter circuit separates the luminance (brightness) and chroma (color) video information, eliminating cross-color that can occur in other sections of the chassis.

*contrast*  The depth of difference between light and dark portions of a TV scene. Also the name given to the control for adjusting the contrast level.

*convergence*  The system that brings the three electron beams together in a color picture tube so that they all pass through the same hole in the shadow mask and strike the correct dots on the screen.

*converter*  A stage in the tuner or front end of a TV set or any radio receiver that converts an incoming signal to a predetermined frequency, called the intermediate frequency (IF). All incoming signals are converted to the same IF frequency.

*corona*  Similar to an electric arc, except that this is a characteristic of much voltages (thousands). Corona occurs as a continuous, fine electrical path through the air between two points, sometimes accompanied by a faint violet glow, usually near the picture tube.

*crystal*  A quartz of synthetic mineral-like slab or wafer having the property of vibrating at a precise rate or frequency. Each crystal is cut to vibrate at the desired frequency. Such a crystal is used in the 3.58-MHz oscillator to control its frequency.

*CRT*  Cathode-ray tube; another name for the color picture tube.

*damper*  A diode, tube, or semiconductor used in horizontal amplifier circuits to suppress certain electrical activity. It, incidentally, provides B+ boost voltage.

*dc*  Abbreviation for direct current.

*deflection*  The orderly movement of the electron beam in a picture (cathode-ray) tube. Horizontal deflection pertains to the left-right movement; vertical deflection is the up-down movement of the beam.

*deflection IC*  Today, the deflection circuits have both the vertical and horizontal oscillator and amplifier circuits in one IC. You might find the deflection circuits in one large IC with many different circuits.

*degaussing*  Demagnetizing. In color TVs, an internal or external circuit device that prevents or corrects any stray magnetization of the iron in the picture-tube faceplate structure. Magnetization results in color distortion.

*demodulator*  A demodulator separates or extracts the desired signal, such as sound energy or picture information from its carrier.

*detector*  Same as demodulator.

*digital multimeter (DMM)*  The digital multimeter can measure voltage, resistance, current, and test diodes. Most DMMs have an LCD display. Today, you can find that the DMM also measures capacity, frequency, tests transistors, and is a frequency counter besides the regular testing features.

*diode*  A two-element electron device: a tube or semiconductor. The simplest and most common application of a diode is in the conversion of ac to dc (rectification).

*electrolytic capacitor*  These capacitors can be used as filter or decoupling capacitors in the TV. Large filter capacitors are used in the low-voltage power supply.

*faceplate*  The front assembly of a picture tube. In a color tube, it includes the tricolor phosphor and the aperture mask.

*field*  One scanning of the scene on the face of the picture tube, in which every alternate line is (temporarily) left blank. The scan duration of a field is ⅟₆₀ second. Two fields, the second one filling in the blank lines left by the first one, make up a frame, or a complete picture. A frame duration is ⅟₃₀ second.

*filter*  The electrolytic filter capacitor is found in the low-voltage power supply. Always replace it with one that has the same voltage and capacitance or higher (never lower values).

*flyback, retrace*  The name given to return movement of the electron beam in a picture tube after completing each line and each field. You don't see flyback or retrace lines (normally) on the picture tube because they are blanked out.

*flyback transformer*  Another name for the horizontal output transformer. The flyback transformer takes the sweep signal from the horizontal output transistor and builds up the high voltage to be rectified for the HV of CRT. The flyback provides horizontal sweep for the yoke circuits.

*focus*  Some picture tubes are constructed internally with self-focusing elements. In other TVs, a focus control varies the voltage applied to the picture-tube focus element. This voltage can vary from 4 to 5.3 kV.

*frame*   The combination of two interlaced fields is called a *frame*. Because it consists of two fields, each of $\frac{1}{60}$ second duration, the frame duration is $\frac{1}{30}$ second.

*frequency*   The number of recurring alternations in an electrical wave, such as ac, radio waves, etc. The frequency is specified by the number of alternations occurring during 1 second and given in hertz (cycles per second), kilohertz (1000 cycles) and megahertz (million cycles).

*frequency counter*   Actually, the frequency counter test instruments count the frequency of various circuits. The frequency range can vary from 2 Hz up to 100 MHz.

*gain*   Relative amplification. The number of times a signal increases in size (level) due to the action of one or more amplifiers. The overall gain of a signal is often millions of times.

*gas*   Refers to the presence (undesirable) of a trace of gas inside a vacuum tube. A gassy tube is a defective tube.

*ghost*   Most commonly a double-exposure type of a scene on the TV screen. Usually a fainter picture appears somewhat offset to the right of the main image caused by the reception of two signals from the same station; one signal is delayed in time.

*G-Y*   The green color signal minus the monochrome.

*high voltage*   Generally refers to the multithousand picture tube voltage, but it can be used to mean any potential of a few hundred volts or more.

*high-voltage probe*   The high-voltage probe is a test instrument that will check the anode and focus high voltage at the CRT. The new probes may measure up to 40,000 Vdc.

*horizontal*   Pertaining to any of the functions associated with left-to-right scanning in a picture tube including the horizontal amplifier, oscillator, frequency, drive, lock.

*HOT*   Horizontal output transformer, which steps up the low-oscillator voltage, usually with a driver and horizontal output transistor between. This voltage is rectified by the HV rectifier and applied to the anode terminal of the picture tube.

*hue*   In color TV, the basic color characteristic that distinguishes red from green from blue, etc.

*hum*   Same as ac hum.

IC (integrated circuit)   A structure similar to a module, in which a number of parts required for the performance of a complete function are prewired and sealed. It is not repairable.

*IF (intermediate frequency)*   In the tuner of a TV or radio receiver, the incoming from the desired station is mixed with a locally generated signal to produce an intermediate signal, usually lower than the frequency of the incoming signal. The IF is the same for all stations. The tuner changes to accommodate each incoming signal.

*IHVT*   The integrated horizontal or high-voltage output transformer has HV diodes and capacitors molded inside the flyback winding area. The new IHVT transformers can also provide several different voltage sources for the TV circuits.

*in-line picture tubes*   A more recent development in color tube structure that produces the three basic colors in adjacent strips or bars, instead of the earlier types, which produced three-dot or triad groups. Improved color quality, as well as simplified design and maintenance, is claimed for this type of design.

*isolation transformer*   The isolation transformer can be a variable type that raises or lowers the power-line voltage to the TV. Always use an isolation transformer with an ac/dc-powered TV chassis.

*intercarrier*   A term describing the current system of TV receiver design in which a common IF system is used both for picture and sound information. In older TVs, the design was split-sound, in which separate IF channels for the picture and sound were used.

*leakage*   Undesired current flow through a component.

*linearity*   Picture symmetry. Horizontal linearity pertains to symmetry between the right and left sides of the picture, best observed with a standard test pattern. Also, an adjustment for achieving such linearity. *Vertical linearity* refers to symmetry between upper and lower halves of a picture.

*line filter*   A device sometimes used between the ac wall outlet and a radio or TV to reduce or eliminate electrical noises.

*line, transmission*   The antenna lead-in wire or cable.

*lock, horizontal*   An adjustment in some TVs for setting the automatic frequency operation on the horizontal sweep oscillator.

*loss*   Usually refers to the amount of signal loss in the antenna lead-in (transmission line). This is particularly serious on UHF.

*low-voltage regulator*   The low-voltage regulator is used in the low-voltage power supply. The regulator can be transistors or an IC. The fixed regulator supplies a well-filtered, regulated, constant voltage.

*microcompressor*   The microcompressor or microcomputer chip is built like a regular IC with 8 to 80 (or more) separate terminals. The microcompressor IC can also have surface-mounted terminals.

*modulation*   The process of combining (by superimposition) a sound or picture signal with a carrier signal for the purposes of efficient transmission through air. The carrier's only function is to piggyback the intelligence.

*module*   A subassembly of a number of parts, usually including transistors and diodes. It is encapsulated and not repairable. See *IC*.

*modular chassis*   A TV chassis that consists entirely of separate modules for each circuit in the TV.

*motor boating*   A "putt-putt" sound caused in the audio sound input and output circuits. Motor-boating can be caused by poorly grounded or poorly filtered circuits.

*oscillator*   Generator of a signal, such as the 3.58-MHz color subcarrier signal, the RF oscillator in the tuner, the horizontal oscillator signal (15,750 Hz) and vertical oscillator signal (60 Hz).

*oscilloscope*   A test instrument that can show exact waveforms throughout the TV circuits to help troubleshooting and locate defective components for the electronic technician.

*PC board*   A subassembly of various parts, not necessarily all for one and the same function, on a phenolic or fiberglass board on which the interconnections are printed on metal veins or paths. No conventional wiring, except external interconnections, is used.

*parallel*   A method of circuit component connection, where all components involved connect to common points so that each component is independent of all other components. For example, all light bulbs in your house are connected in parallel.

*phosphor*   The coating on the interior of the faceplate of a picture tube, which emits light when struck by an electron beam. The chemical composition of the phosphor determines the color of the light it will emit.

*picture projection*   Three small projection color tubes are used to project the TV image on the front or rear of a large screen TV receiver. The projection tubes are found inside the cabinet of a rear projection color set.

*picture tube*   The picture tube receives the video color signal that displays the picture upon the picture tube raster. The new picture tube sizes are 32 and 35-inch.

*power supply*   That portion of a piece of electronic equipment which provides operating voltages for its tubes, ICs, transistors, etc.

*preamplifier*   A high-gain amplifier used to build up a signal so it is strong enough to present to the normal level amplifiers, for example, an antenna preamplifier for fringe area reception.

*pulse*   A single signal of very short duration used for timing and sync purposes. Sync pulses are the best example of this type of signal. Pulses occur in precisely measured bursts.

*purity, color*   The display of the various true colors without any accidental or unwanted contamination of one color by any of the others. Color purity is largely dependent on correct convergence adjustments.

*raster*   The illuminated picture tube screen fully scanned with or without video.

*regulator*   A transistor or IC that regulates the voltage for a given circuit in the low-voltage or HV power supplies.

*remote control*   A hand-held transmitter that controls the function of the TV by the operator from a distance. Today, the stations are tuned in electronically instead of the old method of rotating the tuner with a motor.

*resistance*   Electrical friction represented by the letter R. The ohm is the unit of resistance. Resistance limits current flow.

*RF*   Abbreviation for radio frequency.

*retrace*   The return movement of the scanning electron beam from the extreme right to the extreme left and from the bottom to the top of the raster. Also see flyback.

retrace blanking   The extinction or darkening of the light on the face of the picture tube during retrace time to make these lines invisible. Should retrace blanking fail white lines sloping downward from right to left would be seen on the screen.

*R-Y*   The red color component of the overall color picture signal minus the monochrome.

*sand castle*   The sand-castle generator is a three-level signal pulse that includes horizontal and vertical blanking and burst keying pulses.

*saturation*   Pertains to the full depth of a color, in contrast to a faint, feeble color. Saturated colors are strong colors.

*scanning lines*   The horizontal lines that you can see up close in the picture or raster. The scanning lines make up the picture from left to right, looking at the front of the TV screen.

*SCR*   The silicon-controlled rectifier is used in the low- and high-voltage regulator power-supply circuits. In some TVs, an SCR can be used as the horizontal output transistor.

*semiconductor*   A general name given to transistors, diodes and similar devices in differentiation from vacuum-tube devices.

*series*   A connection between a number of components or tubes in chain fashion (e.g., one component follows the other). If any one component opens or burns out, it breaks the series circuit.

*shadow mask*   Same as aperture mask.

*shield*  A metallic enclosure or container surrounding a component, tube, cable, etc. Also see tube shield.

*shielded cable*  A wire with a metal casing on the outside to prevent unwanted electrical energy from reaching the inner conductor.

*signal*  Electrical energy containing intelligence, such as speech, music, pictures, etc.

*signal-to-noise ratio*  A mathematic expression that indicates the relative strength of a signal within its noise environment. A good signal has a high signal-to-noise ratio.

*solid-state*  A term indicating that the radio, TV, etc., uses semiconductors and not vacuum tubes, but transistors, diodes, etc.

*sound bars*  Thick horizontal lines or bars, usually alternately dark and light, appearing on the TV picture screen due to unwanted sound energy reaching the picture tube. In appearance, the width, number, and position of these bars varies with the nature of the sound. Sound bars are caused by a misadjusted circuit.

*subcarrier*  The color picture information carrier. It is called a *subcarrier* because it is a secondary carrier in the particular channel. The color subcarrier frequency is 3.58 MHz.

*surface-mounted components*  The surface-mounted parts are soldered into the circuit on the same side as the PC wiring. You might find surface-mounted components mounted under the PC chassis with larger components on top in the latest TVs.

*sync*  An abbreviation for a synchronizing signal. It is a timing signal or series of pulses sent by the transmitter and used by the receiver to stay in precise step with the transmitter.

*sync clipper*  See *clipper*.

*sync separator*  A circuit in a TV that separates the sync from the picture information or the vertical sync pulses from the horizontal sync pulses.

*transistor*  A solid-state semiconductor used in amplifier, oscillator, and power circuits of the TV chassis. The transistor operates at lower voltage than the vacuum tube. Some chassis might have both NPN-type and PNP-type transistors.

*trap*  An electrical circuit that absorbs or contains a particular electrical signal (also called *wave*)

*triad*  The three-color, three-dot group (red, green, and blue) of which the color picture tube phosphor is made. Each group of three dots is a triad; thousands of triads are contained on a modern color tube screen.

*triac*  A solid-state controller device usually located in the low-voltage power-supply circuits.

*tripler*  A solid-state component consisting of capacitors and diodes to triple the applied RF voltage from the flyback or horizontal output transformer. In the latest TV, the horizontal output transformer and the high-voltage rectifiers can be molded into one component.

*tuner*  The tuner picks up each broadcast TV signal and passes it to the IF circuits for amplification. The tuner can be operated manually or with a remote control.

*UHF (ultrahigh frequencies)*  Radio and TV frequencies from 300 MHz upward. Channels 14 through 83 are all located in the UHF band and are, therefore, called *UHF stations*.

*varactor*  A semiconductor device with the characteristics of a tunable device through the application of a voltage. In contrast to the conventional frequency variation through the use of coil and capacitor techniques, the varactor require only a voltage variation to effect tuning. In some recent TVs, varactor tuners have been used to replace the conventional coil-switching tuners. Simplicity, greater stability and freedom from deterioration are claimed for this type of tuner.

*vertical*   Pertaining to the circuits and functions associated with the up-down motion or deflection of the electron beam.

*vertical amplifier*   An amplifier following the vertical oscillator used to enlarge the vertical sweep signal.

*VHF (very high frequencies)*   Radio and TV stations located below 300 MHz (down to 50 MHz). TV channels 2 through 13 as well as the FM band are in the VHF spectrum.

*video*   A term applied to picture signals or information (video, circuits, video amplifier, etc.).

*VOM (volt-ohmmeter)*   The first pocket-sized VOMs was used for continuity, voltage resistance, and current tests. The VOM utilizes a meter to display the measured readings.

*wave*   The name given to each recurring variation in alternating electric energy, including radio and TV signals. Also called *analog signal.*

*width*   The width of a TV screen may be pulled in at each side, indicating problems within the horizontal deflection system. Poor width can be caused by the HV regulator transistors, SCRs, and zener diodes in the regulator circuit. Poorly soldered pincushion transformer connections can cause width problems.

# Computers, Printers, Copiers, and Fax Machines

Use this glossary section in conjunction with the chapters on computers, printers, copiers and fax machines.

*AA (Automatic Adjust)*   The document setting you use for sending documents, such as text and photos.

*ac (alternating current)*   The type of electrical current available from a wall outlet.

*activity report*   Journal of transactions, both sent and received.

*ADF*   Automatic Document Feeder

*automatic dialing*   Dialing fax or telephone numbers by pressing one or three buttons.

*ANS hook up/Manual*   The light and button that indicate and control how the fax machine detects whether a call is from a fax or a telephone.

*Auto Fax/Tel Switching*   See *FAX/TEL switching.*

*baud rate*   See *Sending speed.*

*bidirectional parallel interface port*   An interface connection that is capable of both sending or receiving. For example, when you print and when you send a fax from your computer, data goes from your computer to the printer.

*bidirectional printing*   The ability of the fax machine to print both left to right and right to left. This printing method provides a fast speed. See also *Unidirectional printing.*

*bps (bits per second)*   Refers to the speed with which a fax machine sends and receives data.

*broadcasting*   Transmitting documents to more than one location.

*bubble-jet printing*   An ink-jet type printing method that heats the ink to a boiling point to form a bubble. When the bubble expands, no room is left in the nozzle for the ink and the ink is projected onto the paper.

*coded speed dialing*  An automatic dialing method that allows you to dial a fax or telephone number by pressing three buttons: the Coded Dial/Directory button and a two-digit code using the numeric keypad.

*confidential mailbox number*  Two-digit numbers between 00 and 99 used to arrange for confidential sending of documents. See *confidential sending*.

*confidential sending*  The ability to send a document confidentially. The receiving fax machine will keep the document in memory until the intended recipient enters a two-digit code to print the document.

*cursor*  The underline symbol you see on the LCD display when you register numbers and names on some printers. Press the arrow buttons to move the cursor.

*default*  The preset value or factory setting used when you do not set a different one. You can change default values by using the Function button to access the menu system.

*delayed transmission*  The ability to send a document at a preset time in the future. You do not have to be in your office to use delayed sending to one or more destinations.

*density control*  A setting that darkens or lightens the scanning of documents.

*dialing methods*  Ways of pressing one or more buttons to access a number to connect to an outside party or fax machine. Dialing methods include one-touch, coded speed dialing, group dialing, directory dialing, and manual (regular).

*direct sending*  Transmitting a fax document one page at a time without having the document scanned into memory.

*directory dialing*  A dialing method that allows you to dial any telephone or fax number registered for one-touch or coded-speed dialing. You recall the number by the name you entered when registering the number.

*document*  The sheet of paper containing the data that you send to or receive from a fax.

*dpi (dots per inch)*  A unit of measurement for indicating a printer's resolution.

*DRAPED (Distinctive Ringing Pattern Detector)*  Allows you to assign up to five different ring patterns to distinguish voice and fax calls using your telephone company's special services.

*emulation*  A technique where one device imitates (acts like) another device.

*expanded dialing*  Ability to register and then dial a fax or telephone number up to 118-digits long by pressing just one or three buttons.

*extension*  A telephone connected to the fax machine that is used in place of the handset. You can use the extension telephone to activate incoming reception of documents manually.

*FAX/TEL switching*  This option allows you to set the fax machine to automatically detect whether a call is from a fax or a telephone. If the call is from another fax, the transmission is automatically received. If the call is from a telephone, the fax rings to let you know, so you can pick up the handset. With this feature, one telephone line can be shared by both the telephone and the fax.

*FINE*  The setting for documents with very small characters and lines.

*G3, Group 3 fax machine*  Defined by CCITT. Uses encoding schemes to transmit image data while reducing the amount of data that needs to be transmitted, thus reducing transmission time. G3 fax machines can transmit one page in less than one minute. Encoding schemes for G3 fax machines are Modified Huffman (MH), Modified Read (MR), and Modified Modified Read (MMR).

*group dialing*   A dialing method that enables you to dial up to 95 registered one-touch speed dialing or coded-speed dialing numbers together as a group. This means that you can press just one or three buttons to send the same document to many destinations.

*Halftone*   The document setting used to send documents with intermediate tones, such as photographs.

*HOOK*   The button that engages and disengages the telephone line.

*ink cartridge*   The special type of ink cartridge used with bubble-jet printers.

*Ink detector*   The printer setting that allows you to check if there is enough ink in the ink cartridge. This option prints a small box in the bottom right corner of incoming documents.

*interface*   The connection between two devices that makes it possible for them to communicate with each other.

*interface port*   This is usually an 8-bit, bidirectional parallel interface port.

*jacks*   The telephone receptacles on your wall or in your fax machine used to connect it to the telephone line, handset, answering machine, extension telephone, or data modem.

*manual dialing*   Pressing the individual buttons on the numeric keypad to dial a fax or telephone number. Also called *regular dialing*.

*Manual receiving*   A setting that allows you to answer all incoming telephone and fax calls. A slow beep indicates an incoming fax transmission from another machine. Just push the Start/Copy button to receive the incoming fax.

*manual redialing*   When you use regular dialing, you can redial a number manually simply by pressing Redial on the operation panel. The last number called is the number redialed.

*memory broadcast*   The ability to scan documents into memory and send it to as many as 97 locations using automatic or manual dialing. If you use this feature regularly for the same locations, see *Group dialing*.

*memory sending*   Scans a document into memory before the MultiPass 1000 dials the number and sends it. This method is faster than direct sending and it allows you to retrieve your original document immediately after scanning.

*modem*   A device that converts (*mo*dulates) digital data from transmission over telephone lines. At the receiving end, this device converts the modulated data (*dem*odulates to digital format that the computer understands.

*noise*   A term applied to a variety of problems that impair the operation of telephone lines used for fax and modem communication.

*numeric keypad*   The round numbered buttons on the operation panel marked the same as a standard telephone keypad. Press them to perform manual dialing. You also use the numeric buttons to enter numbers and letters when you register numbers and names, and for entering coded speed dialing or confidential sending codes with two or more digits.

*one-touch speed dialing keypad*   The rectangular buttons numbered 01 to 16 on the operation panel, each of which can be registered as a fax and/or telephone number. Once a number is registered, you press one button to dial the entire number.

*pause*   A timing entry required for registering certain long-distance numbers and for dialing out through some telephone systems or switchboards.

*PBX (private branch exchange)*   See *Switchboard*.

*polling*   Requesting another fax machine to send a document to your fax machine. This feature is useful for obtaining a document when the original document is waiting in the other fax and the fax operator is not there. See *polling ID*.

*polling ID* An eight-digit binary number (binary = 0 or 1) used to control your ability to request another fax machine to send documents to your fax machine. The polling ID you register must match the ID used in the polling network. See *polling*.

*pulse* See *Rotary pulse*.

*RAM (random-access memory)* Memory that is used for temporary storage of information, such as your scanned or received documents.

*redialing, automatic* When the fax you dial does not answer or a sending error occurs, some fax machines wait for a specified interval and then redials the same fax number. You can adjust the number of redials and the length of time between redialing.

*redialing, manual* When you use the regular dialing method, you can quickly call the last number dialed by pressing the Redial button.

*reduction mode* An automatic feature that slightly reduces the received image to allow room at the top of the page for the sender's ID information. You can also reduce the size of large incoming documents using the RX Reduction option.

*registering* A process by which you place fax or telephone numbers and names in the fax machine's memory for automatic dialing so that you can save time dialing frequently called destinations.

*regular dialing* See *manual dialing*.

*relay broadcast* See *relay sending*.

*relay sending* Transmitting a document to more than one location through another relay fax. This is cost effective if you want to send a document long distance to a group of offices located in the same area. Sending the document directly to each office would require one long distance call per document; sending the document through a relay fax would require one long distance call and the relay fax would make local or short toll calls to send to the nearby destinations. Also called *relay broadcast*.

*remote receiving ID* The two-digit code that enables you to manually activate a fax using an extension telephone that is connected to your fax machine.

*remote reception* Activating a fax by answering an extension telephone that is connected to, but that is not located near.

*resolution* The density of dots for any given output device. Expressed in terms of dots per inch (dpi). Low resolution causes font characters and graphics to have a jagged appearance. Higher resolution means smoother curves and angles as well as a better match to traditional typeface designs. Resolution values are represented by horizontal data and vertical data, for example, $360 \times 360$ dpi.

*rotary pulse* A telephone dialing system where a dial is rotated to send pulses to the telephone switching system. When you pulse dial, you hear clicks. When you Touch-Tone dial, the most common dialing system, you hear tones. Rotary pulse dialing requires certain setting adjustments.

*sender ID* The identifying information from the sender at the top of a document including: date and time, the sender's fax/telephone number, sender's name, receiver's name or company name, and page number. Also called *TTI* or *Transmit Terminal ID*.

*sending speed* The bits-per-second rate at which documents are sent. See also *bps*.

*Standard* A document setting for sending normal typewritten or printed documents containing only text and no drawings, photographs, or illustrations.

*Standby* The mode in which the fax machine is on and ready to use. All operations start from Standby mode when the LCD displays the date and time.

*switchboard*    Also called a *PBX (Private Branch Exchange)* internal switching system. A telephone system, usually for a large company office with many extensions, whereby you must dial an outside line number along with the regular telephone or fax number. Dialing out through a switchboard sometimes requires use of the Pause button.

*timed sending*    Setting the fax machine to transmit documents at a preset time in the future. See also *Delayed transmission.*

*tone/pulse setting*    The ability to set the fax machine to match the telephone dialing system your telephone line uses: Touch-Tone or rotary pulse.

*Tone*    A button that allows you to temporarily switch to Touch-Tone from pulse dialing. In some countries, on-line data services might require that you use tone dialing.

*TTI Transmit Terminal ID*    See *Sender ID.* A protocol and an application programmer's interface (API) that allows you to input image data directly from any source (for example: desktop and handhold scanners, video-capture boards, digital cameras, and other imaging equipment) without requiring users to switch out of the application. It provides compatibility between image input devices and applications by acting as the liaison between hardware devices and software applications.

*unidirectional printing*    Printing in one direction only, left to right. This printing method provides a higher image quality, but slower print speed. See also *bidirectional printing.*

# INDEX

## ABOUT THE AUTHOR

**Robert L. Goodman** is one of the nation's most popular and esteemed electronics writers. The author of over 60 books on practical electronics, he wrote his first color TV service manual in 1968 for TAB Electronics. Among his recent books are *Troubleshooting and Repairing Color Television Systems*, *Digital Satellite Service*, and *Maintaining and Repairing VCRs*. A number of his books have been translated into foreign languages, including Chinese, and are international bestsellers. Bob not only writes about electronics, he also works on his own electronics bench and in his own electronics lab. A developer of electronic testing equipment, he writes his books on an office deck overlooking the Quachita Mountains and lakes of western Arkansas.